Supernovae Spectra
(La Jolla Institute, 1980)

AIP Conference Proceedings
Series Editor: Hugh C. Wolfe
Number 63

Supernovae Spectra

(La Jolla Institute, 1980)

Editors

Roland Meyerott and George H. Gillespie
La Jolla Institute

American Institute of Physics
New York   1980

L.C. Catalog Card No. 80-70019
ISBN 0-88318-162-2
DOE CONF- 800126

V.

PREFACE

The La Jolla Institute Workshop on Atomic Physics and Spectroscopy for Supernovae Spectra, held January 10-12, 1980, brought active researchers on supernovae together with workers in the areas of atomic physics and spectroscopy. The Workshop included talks by invited speakers on the observed supernovae spectra, plasma conditions in supernovae envelopes, on the present status and future requirements of atomic physics and spectroscopy that contribute to the solutions of the supernovae problems. One day was devoted to informal discussions in small working groups. Several contributed papers were presented at the close of the meeting.

We are indebted to the participants for contributing the additional effort to make this interdisciplinary workshop a success. Direct communication links between researchers on supernovae and atomic physics and spectroscopy were established at the meeting and information has already started to be exchanged. It is hoped that this process will continue, as more atomic physics and spectroscopic results of interest to the understanding of supernovae become available.

Finally, we are very grateful to Adrienne Ott, who assembled this volume and coordinated the La Jolla Conference, to Adolf R. Hochstim, whose support and encouragement made the whole show possible, to the La Jolla Institute for sponsoring and supporting the Workshop, and to the National Science Foundation for partial support.

Roland E. Meyerott

George H. Gillespie

La Jolla Institute
La Jolla, California

TABLE OF CONTENTS

VII.

DENSITY, VELOCITY, AND TEMPERATURE PROFILES FOR THE EXTENDED ENVELOPE MODEL OF TYPE I SUPERNOVAE

Gordon Lasher[*]

IBM - T. J. Watson Research Center

Yorktown Heights, NY 10598

ABSTRACT

The early light curve of Type I Supernova has been fitted by a model in which the low density envelope of a supergiant is exploded by the sudden release of energy at its center. This paper supplements previous publications on this model by giving information about the model which is necessary for the computation of the emitted spectrum, namely, the density, temperature and velocity profiles of the expanding shell.

INTRODUCTION

The first detailed model of Type I supernova was that of Colgate and McKee[1] who, following a suggestion of Truran's, computed the light emission from an exploding white dwarf containing large amounts of Ni^{56}. For a discussion of recent work on this model see these proceedings and Chevalier's review.[2]

In 1973 Barbon, Ciatti and Rosino[3] published an average light curve for Type I supernovae demonstrating that they form a quite uniform set of phenomena. Stimulated by the work of C. Gordon,[4] I constructed a very simple model in which the sudden deposition of energy at the center of a low density supergiant stellar envelope produced a light curve which could be fitted to the average observed light curve of Barbon et al. This work is reported in reference 5.

In an expanding medium the effective opacity can be much greater than in a stationary one because the photons are continually red shifted and have a greater probability of interacting with the narrow transitions between discrete atomic energy states. Karp, Lasher, Chan, and Salpeter[6] analyzed this effect. They showed that in an optically thick region it could be taken into account by defining an effective opacity, called the expansion opacity, and they computed this quantity for a number of cases.

The expansion opacity was used to investigate the effect of envelope composition in reference 7. The most striking result of that work was the distortion of the light curve by a cooling or deionization wave when the stellar envelope contains very little hydrogen or other material with a low ionization potential.

There is no generally accepted evolutionary track that leads to the kind of progenitor star required by the Type I supergiant model. C. Wheeler,[8] however, has proposed that R CoBor stars may become Type I supernovae, and K. Nomoto has computed an evolutionary track which ends in a helium supergiant similar to this model.[9]

[*]Address for January-May 1980: Physics Dept., Stanford University, Stanford CA 94350.

ISSN:0094-243X/80/630001-06$1.50 Copyright 1980 American Institute of Physics

THE SPECIFICATIONS OF THE MODEL

The results given in this paper are from a program that is essentially the same as that of the original supergiant Type I models of reference 5. The starting point is a stationary sphere of gas representing the envelope of the progenitor star. At time zero certain amount of energy is injected into the center of the envelope presumably by the collapse of the degenerate core whose free-fall time is of the order a second. Then the equations of one-dimensional, spherical hydrodynamic flow and grey radiation diffusion are integrated to compute the outgoing shockwave and the subsequent inward rarefaction wave; both of which accelerate the envelope in the outward direction and lead to the formation of a supernova shell with velocities of the order of 10^9 cm/s. When the shell's optical thickness is reduced sufficiently, the diffusion of radiation out of the shell becomes significant. The maximum luminosity occurs when the shell is 20 or 30 radiation lengths thick.

The opacity in the relevant region of temperature and density is due mostly to Thompson scattering with a modest correction due to the expansion opacity of Karp et al.[6] In the model here the opacity was taken to be constant and equal to 3.48 cm^2/g. The expansion opacity is important not only in making more precise computations, but also in providing an effective mechanism for keeping the radiation in thermal equilibrium as it diffuses out of the expanding shell.

In reference 5 the parameters of the model are adjusted to give the observed values of the maximum luminosity, the time duration, and expansion velocity. This process determines the energy of the explosion and the initial density and mass of the stellar envelope. A model with an energy of 10^{51} erg, and a density and mass of 10^{-8} g/cm^3 and 2 solar masses gives a good fit. The light curve is not very sensitive to the initial density and that parameter is only determined to within a factor of ten.

THE STRUCTURE OF THE EXPANDING SHELL

The structure of the supernova shell is given in the three figures. There are two models, one with a stellar envelope of constant density and one whose envelope density decreases by a factor of ten from the center to the outer edge. These two models give almost identical light curves, but Fig. 1. shows us that the density structure of the supernova shell is very sensitive to the density structure of the stellar envelope.

After about 50 days the pressure in the shell is too low to give it any significant further acceleration. For this reason the density data is given at a late time, namely at 100 days, and may be scaled by proper powers of the time to obtain the density versus radius at late times. With an error of about 20% one can use the scaled densities as early as 25 days.

In a two dimensional calculation Chevalier and Klein[10] found that Rayleigh-Taylor instability resulted in the expanding shell forming strong clumps of material with about half the shell's mass in the clumps. This effect would effectively smear out the thin, high density shell of Fig. 1. On the other hand the steep density gradients of Fig. 1 would lead to strong clumping of the shell on scales much smaller than the computation of Chevalier and Klein could deal with.

REFERENCES

1. S. A. Colgate and C. McKee, Ap. J. *157*, 623 (1969).
2. R. A. Chevalier, preprint of a chapter in *Fundamentals of Cosmic Physics*.
3. R. Barbon, F. Ciatti, and L. Rosino, Astr. and Ap. 25, *241* (1973).
4. C. Gordon, Ap. J. *207*, 860 (1976).
5. G. Lasher, Ap. J. *201*, 194 (1975).
6. A. H. Karp, G. Lasher, K. L. Chan and E. E. Salpeter, Ap. J. *214*, 161 (1977).
7. G. Lasher, A. H. Karp, and K. L. Chan in *Supernovae*, D. N. Schramm, ed. (1977).
8. J. C. Wheeler, Ap. J. *225*, 212 (1978).
9. K. Nomoto, these proceedings (1980).
10. R. A. Chevalier and R. I. Klein, Ap. J. *219*, 994 (1978).

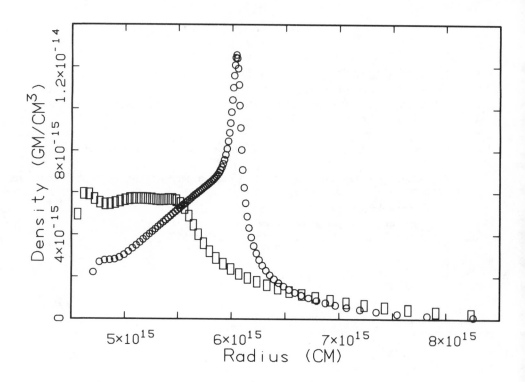

Fig. 1. The mass density versus radius profiles for two models with an explosive energy of 10^{51} erg released within a 2 solar mass envelope having an initial average density of 10^{-8} g/cm^3. The highly peaked curve whose points are circles is from a model with an envelope of constant density while the one plotted by rectangles is from an envelope whose innermost material has an initial density ten times that of the material at its outer edge. The two points corresponding to the two outermost zones have not been plotted for this model; they lie to the right of the end of the radius scale. The data was taken at 100 days after the explosion when the velocity is nearly at its final value. The density structure for other times can therefore be found by properly scaling the values of this figure. The error in this scaling is about 20% at 25 days and less than 1% at 50 days.

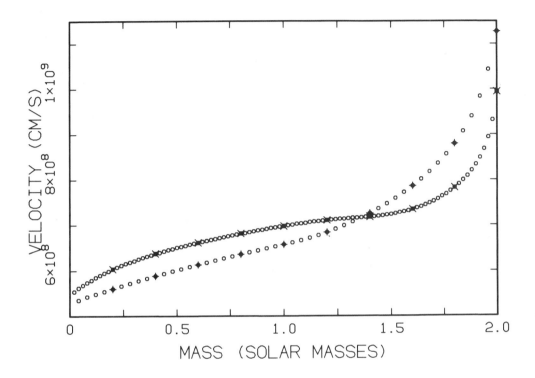

Fig. 2. The final velocity versus mass coordinate of the two models of Fig. 1. The model with a constant initial density has the higher velocity near the center, has 100 zones, and every tenth zone is marked '×'; while the model whose initial density decreases with radius, has 50 zones, and every fifth zone is marked '+'. The sharp peak of Fig. 1. corresponds to the almost horizontal section of the first curve.

These curves are good approximations to the solution of the corresponding continuous hydrodynamic problem. They clearly extrapolate to a nonzero velocity at zero mass coordinate and therefore indicate that the gas density drops suddenly to zero at the inner edge of the expanding supernova shell. This discontinuity in density is not accompanied by a discontinuity in pressure because the central vacuum is filled with radiation.

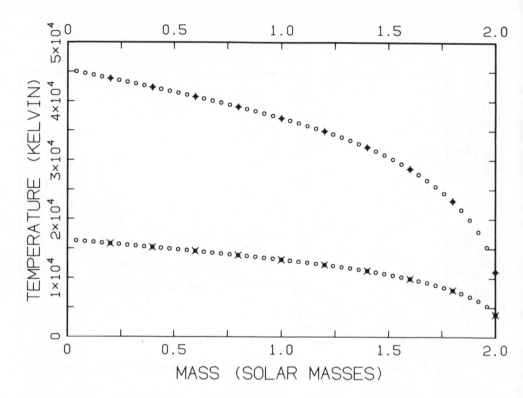

Fig. 3. The temperature versus mass coordinate of the constant density model at two times. The upper curve is at 12.2 days near the time of maximum blue magnitude of -18.43 and the lower curve is at 28.4 days when the blue magnitude has dropped to -17.48.

THE LIGHT CURVE OF TYPE I SUPERNOVAE

S. A. Colgate
University of California, Los Alamos Scientific Laboratory
P.O. Box 1663, Los Alamos, NM 87545
and
New Mexico Institute of Mining and Technology
Socorro, NM 87801

Albert G. Petschek
New Mexico Institute of Mining and Technology
Socorro, NM 87801

John T. Kriese
University of California, Los Alamos Scientific Laboratory
P.O. Box 1663, Los Alamos, NM 87545

ABSTRACT

Calculations of the intermediate and late time luminosity of type I supernovae based on 100% efficiency for optical emission of energy deposited by the Ni^{56} decay chain give good agreement with observations provided $M_{ej} v^{-2} = (2.2 \pm 0.5) \times 10^{17} M_{\theta} s^2 cm^{-2}$ where M_{ej} is the ejected mass and v is the expansion velocity. Account must be taken of the escape of both gamma rays and positrons. These two escape processes as well as the early luminosity peak, as calculated by Colgate and McKee, are all consistent with the same value of M_{ej}/v^2.

INTRODUCTION

Type I supernovae are recognized by a characteristic light curve that has been outlined in several publications. Some of these use a superposition of many supernova light curves (Barbon et al 1973; Morrison and Sartori 1969). The problem with curves determined by superposition of many light curves is the variations that are produced by adjustment of explosion time and normalization of magnitude. It therefore seems preferable to compare a theory of supernova light curves with one or two careful and extensive measurements of individual light curves. Particularly, the older supernova NGC 4182 (Baade and Zwicky 1938; Van Hise 1974) and the more modern supernova NGC 5253 (Kirshner and Oke

1975) seem optimal because of the long period of observation (600 to 700 days) and the high peak luminosity. We will therefore consider these two as our standard.

NGC 4182 has been analyzed in terms of two exponential decays by Van Hise (1974). It was this analysis that caused many to seek an answer to the tantalizing suggestion that a factor of exactly 3/4 was involved in the transformation of the half-lives of a presumed $Ni^{56} \to Co^{56} \to Fe^{56}$ beta decay to the half-lives of the observed luminosity decay. It is the purpose of this paper to show that the apparent luminosity half-lives equal to 3/4 of the radioactive half-lives are an entirely fortuitous consequence of the progressive transparency of the expanding nebula.

We start with the work of Colgate and McKee (1969) who showed that the magnitude and width at half maximum of the early light curve of a type I supernova could be obtained by using the radioactive energy of 0.25 M_Θ of Ni^{56} and accounting for diffusive escape of radiation. These calculations required the ejection of 0.75 M_Θ of silicon burning products at a velocity of 1.5×10^9 cm s^{-1}. The maximum diffusive release of radiant energy occurs at an optical depth $\rho r/\lambda = \tau \cong 3$ c/v where v is expansion velocity of the envelope. Approximately 1/2 the solar mass of matter was required to supply the total energy and opacity. Adiabatic expansion absorbs a fraction, 44%, of the radioactive energy before diffusive release at 6 days. The peak in the supernova light curve corresponds to 10^{43} ergs/s or a bolometric magnitude of 20 in the blue with no correction. These are the numbers one would obtain for the peak luminosity of type I supernova for a Hubble constant of 50. On the other hand, the larger Hubble constant of 100 currently being discussed would imply that the total luminosity would be closer to 4×10^{43} ergs/s and that 1 M_Θ of Ni^{56} would have to be ejected. Furthermore, the transparency function that we are about to derive implies that the total kinetic energy of the ejected matter would have to be 2.7×10^{51} ergs and this would only make sense if one of the combination detonation models of an external helium envelope and an internal carbon-oxygen core suggested by Weaver and Woolsey (1980) were to be the cause of type I supernova.

DEPOSITION CALCULATIONS

The gamma-ray Monte Carlo photon transport code "MCP" (Cashwell et al 1973) of the Theoretical Design Division of Los Alamos was used to calculate the gamma ray deposition energy as a function of time for the gamma ray spectrum of the $Ni^{56} \rightarrow Co^{56} \rightarrow Fe^{56}$ decay scheme. The gamma ray absorption mean free path λ is adequately represented by a constant 35.5 g cm^{-2} for either the Ni^{56} or the Co^{56} decay spectrum. The detailed calculations confirm that the deposition function is determined by just ρr and is unaffected by the change in spectrum. The fractional deposition as a function of $\tau = \rho r/\lambda$ is given in Table I. An analytical fit is $D = G[1 + 2G(1 - G)(1 - .75G)]$ where $G = \tau/(1.6 + \tau)$.

Table I Deposition Function for a Uniform Sphere

$Ni^{56} \rightarrow Co^{56} \rightarrow Fe^{56}$ gamma rays.			
τ	D	τ	D
16	.965	1	.517
8	.930	½	.301
4	.857	¼	.158
2	.725	1/8	.080
For $\tau \leq$ ¼, D = 0.64 τ; $\tau = \rho r/\lambda$.			

Another model with the radioactive source restricted to the inner ¼ mass and 3/4 of the mass external as H and He gave the same deposition function within a few percent.

Figure 1 shows the product of the transparency function given in Table I and an exponential of half-life t_1. We characterize the transparency function by the time t_0, at which $D \cong$ ½, i.e., when $\tau = 1$. Figure 1 shows the product of the deposition fraction times the exponential for various ratios of t_0/t_1. The dashed line is drawn corresponding to an exponential of time constant 3/4 t_1, i.e., that modification of the exponential inferred from Van Hise's analysis of the early and late time light curves of type I supernovae. It is evident that at $t_0/t_1 = 4$, an approximate exponential with a half-life equal to 3/4 of t_1 results. In the model we are about to describe $t_0 = 20$ days for γ-ray transparency. The Ni^{56} half-life $t_1 = 6.1$ days is roughly ¼ of t_0 (actually, $t_0/t_1 = 3.3$).

In a separate publication (Colgate, Petschek, and Kriese 1980) we give the justification for assuming that the energy deposition of electrons can be treated in the same fashion as the gamma rays but with a mean free path of $\lambda_\beta = 0.10$ g cm^{-2} rather than the gamma ray value of 35.5 g cm^{-2}. This also assumes that either no magnetic field is present or that the magnetic field is combed radially by the ejection of a relativistic mass fraction. The ejection of a relativistic mass fraction of a value necessary to comb a magnetic field radially is entirely consistent with our prior explanation of the relativistic shock ejection mechanism of cosmic ray formation. The energy in the magnetic field is of the order of 10^{39} ergs whereas that in the relativistic ejected mass fraction should be several 10^{49} ergs. In that case, t_0 for transparency to β-rays is about 370 days, about 4 (actually 4.8) times the Co^{56} half-life of 77 days.

The implication is that the observed 4.8-day and 56-day half-lives are fortuitous combinations of a radioactive decay and a transparency function.

RESULTS

Thus, we recalculated the luminosity of type I supernova (Figs 2 and 3) by using (1) the diffusive release of Ni^{56} decay energy (Colgate and McKee 1969); (2) the progressive gamma ray transparency as calculated by the Monte Carlo gamma-ray simulation code and (3) the fractional deposition of positrons (Arnett 1979) using either zero magnetic field or a radially combed dipole field. After the initial black body peak, we assume 100% optical fluorescence efficiency from Fe^+ (Meyerott 1980). The deposition function determined by the Monte Carlo calculations is then applied to the beta energy source as well as the gamma source. When the nebula is one gamma-ray mean free path thick at $t_0 = 20$ days, good agreement with observations is obtained. The luminosity half-life of 56 days requires that M_{ej}

$v_9^{-2} = 0.22 \pm 0.05$ where the ejected mass is in solar masses and v_9 is the expansion velocity in units of 10^9 cm s^{-1}.

KINETIC ENERGY OF EJECTED MATTER

The kinetic energy of the ejected matter (assuming a uniform density nebula) is $3/5\ M_{ej}\ v^2/2$. Using the estimate of M_{ej}/v^2 above this becomes $2.7 \times 10^{51}\ M_{ej}$ ergs where again M_{ej} is measured in solar mass units. When M_{ej} is 0.5 ($v_9 = 1.5$), the ejected kinetic energy is acceptable, 6.7×10^{50} ergs, but if a larger ejected mass is assumed, the energy requirements become severe. The conversion of 30% to 50% of the ejected fraction of a presupernova carbon core to Ni^{56} is possible in silicon burning (Truran et al 1967). Finally the temperature of the black-body phase of the light curve remains constant for the scaling $M_{ej}\ v_9^{-2} =$ constant.

ACKNOWLEDGMENT

We are indebted to Robert Kirshner and Rollie Meyerott for continuing discussions, and particularly to Dave Arnett who made what seemed to us the unlikely suggestion of the importance of positron transparency. This work was supported by DOE and the Astronomy Section of NSF.

REFERENCES

W. D. Arnett, Astrophys. J. 230, L37 (1979).

W. Baade and F. Zwicky, Astrophys. J. 88, 411 (1938).

R. Barbon, F. Ciati, and L. Rosino, Astron. Astrophys. 25, 241 (1973).

E. D. Cashwell, J. R. Neergaard, C. J. Everett, R. G. Schrandt, W. M. Taylor, and G. D. Turner, Los Alamos Scientific Laboratory Report LA-5157-MS (1973).

S. A. Colgate and C. McKee, Astrophys. J. 157, 623 (1969).

S. A. Colgate, A. G. Petschek, and John T. Kriese, submitted to Astrophys. J. Lett. (1980).

R. P. Kirshner and J. B. Oke, Astrophys. J. 200, 574 (1975).

R. E. Meyerott, Astrophys. J. in press (1980); also this conference (1980).

P. Morrison and L. Sartori, Astrophys. J. 158, 541 (1969).

J. W. Truran, W. D. Arnett, and A. G. W. Cameron, Can. J. Phys. 45, 2315 (1967).

J. R. Van Hise, Astrophys. J. 192, 657 (1974).

T. Weaver and S. Woosley, this conference (1980).

12

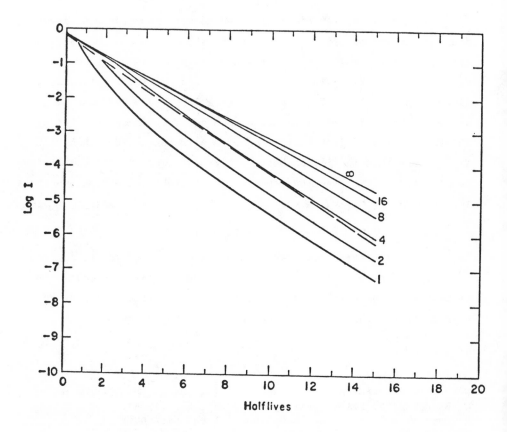

Fig. 1. We plot $D\, e^{-t\ell n2/t_1}$ for various values of t_0/t_1 where t_0 is the time at which deposition is $\frac{1}{2}$. If we assume $M_{ej}/v^2 = \kappa$, then

$$\tau = \rho r/\lambda = \frac{\kappa}{\frac{4}{3}\pi\, t^2\,\lambda}.$$ From Table I $\tau=1$ when $t=t_0$ so that

$$t_0 = \left(\frac{\kappa}{\frac{4}{3}\pi\,\lambda}\right)^{\frac{1}{2}}$$

The deposition D is a function of τ given in Table I. For our best fit model and the early decay of 6.1 days, $t_0/t_1 = 3.3$. In the second part of the decay where $t_1 = 77$ days, $t_0/t_1 = 4.8$.

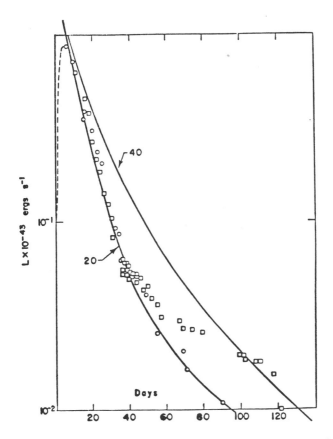

Fig. 2. The calculated luminosity at early and intermediate times for M_{Ni} = 0.25 solar masses and the corresponding deposition functions for $\tau = 1$ at 20 days and 40 days. Gamma ray deposition and the Ni → Co → Fe decay determine the solid curves. The dashed curve is the modification of the deposition function due to diffusion and expansion (Colgate and McKee 1969). The extrapolation of the deposition curves reaches 2×10^{43} ergs s^{-1} at t = 0. The difference between this extrapolation and the dashed curve is due to heat energy converted to kinetic by expansion. The circles, give NGC 5253 data (Kirshner and Oke 1975) and the squares give NGC 4182 (Baade and Zwicky 1938; Van Hise 1974). The circles, NGC 5253, give a better fit. It may be that photometric corrections result in the disagreement of the squares in the interval 50 to 80 days.

14

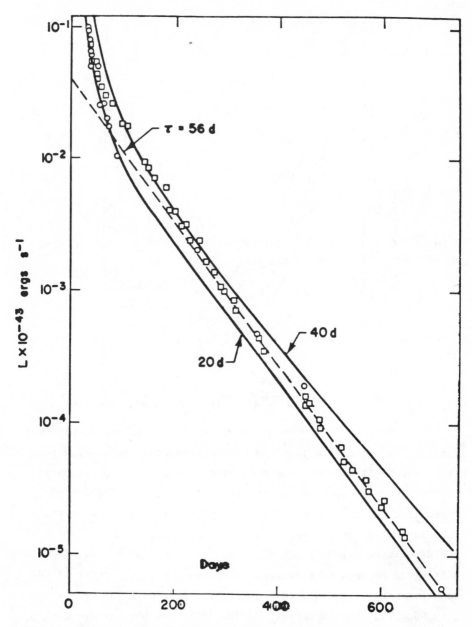

Fig. 3. Same as Fig. 2 for times out to 700 days. Here the
curves are primarily determined by the deposition of positrons
from the Co → Fe decay. The dashed line is a fit to the data with
a slope corresponding to a 56-day half-life.

SUPERNOVA MODELS AND LIGHT CURVES*

Thomas A. Weaver
Lawrence Livermore Laboratory
Livermore, CA 94550

S. E. Woosley[†]
University of California, Santa Cruz
Santa Cruz, CA 95064
and
Lawrence Livermore Laboratory
Livermore, CA 94550

ABSTRACT

This talk briefly reviews the current status of our under-standing of Type II supernovae with particular emphasis on the processes responsible for the emission of electromagnetic radiation. In addition, a relatively novel evolutionary scenario that appears to lead to a Type I supernova explosion is presented.

TYPE II SUPERNOVAE

Type II supernova have long been associated with massive stars with extended hydrogen envelopes. This association is due both to direct spectroscopic evidence[1,2,3] and to the quite distinct correlation of Type II supernovae with the spiral arms of galaxies[4], suggesting that their progenitors are bright, short-lived (<30 million years) O and/or B stars with mass $\gtrsim 10$ $M_\odot$.

We[5-8] have approached this issue from the theoretical viewpoint of evolving massive stars from the zero-age main sequence, through their various hydrostatic and explosive nuclear burning phases, and then comparing the characteristics of the supernova explosions that result to observations. Our numerical model[5] of these events incorporates implicit hydrodynamics with time-dependent convection and semiconvection, and a careful treatment of the complex nuclear processes that characterize the advanced nuclear burning stages. Spherical symmetry, and thus the absence of rotation and magnetic fields, is assumed, as is the unimportance of mass loss.

Complete evolutionary calculations have been performed for 15 and 25 $M_\odot$ Population I stars, and have been summarized by Weaver and Woosley[6]. (See Ref. 5, 6, and 7 for a discussion of the relation of these calculations to previous work.) We shall concentrate in this talk on the observable effects of the explosive death of these stars.

*Work performed under the auspices of the U.S. Department of Energy by LLNL under contract number W-7405-ENG-48.

†Work performed in part under NSF contract number AST-79-09102.

ISSN:0094-243X/80/630015-18$1.50 Copyright 1980 American Institute of Physics

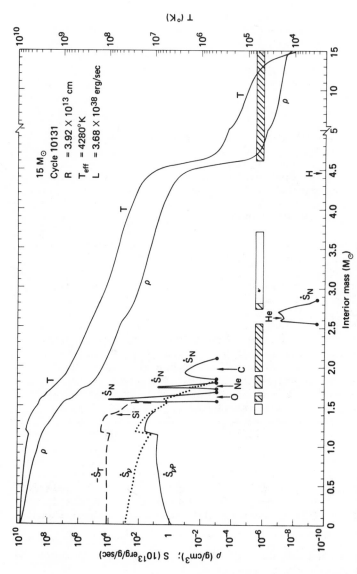

Fig. 1 The thermodynamic structure of the presupernova 15 $M_\odot$ model star is shown as a function of mass fraction.5 Note that the density and temperature profiles are plotted such that the curves will maintain constant separation for the case $\rho \propto T^3$. Here $-\dot{S}_T$ is the total local energy loss rate due to both neutrino emission and nuclear photodisintegration, $\dot{S}_\nu^T$ is the total neutrino energy loss rate, and $\dot{S}_\nu^P$ is the neutrino energy loss rate due to the thermal plasma processes. The nuclear energy generation rate profiles for the various nuclear burning shells are labelled $\dot{S}_N$, and the principal nuclear fuel is indicated. All energy generation and loss rates share the common scale denoted by "S". Active convective regions are indicated by striped bars, while semi-convective and convectively neutral regions are shown as outlined bars. In this figure, R, T_{eff}, and L denote the photospheric radius, effective temperature, and optical luminosity, respectively.

PRESUPERNOVA EVOLUTION

We find that the massive stars we have studied gradually deplete their store of nuclear fuel by burning their initially predominantly hydrogen composition successively into helium, carbon, neon, oxygen, silicon and finally iron. In general, each successive fuel is ignited in the star's core, while shells of the lighter elements remain in the region outside the core, usually separated by active nuclear burning shells. These burning shells are typically convective and are separated by sufficiently steep entropy gradients so that mixing between shells does not occur. The collapse of the core is triggered by the endothermic photodisintegration of iron[5] (c.f. Burbidge, Burbidge, Fowler, and Hoyle[9]). At this stage, about 2/3 of the star's mass resides in a tenuous red-supergiant envelope composed mostly of hydrogen (~60%) and helium (~40%), and having nearly constant $10^{-8} g \ cm^{-1}$ density and 10^5 K temperature (see Fig. 1 and Ref. 5). The radii of the 15 and 25 $M_\odot$ stars are respectively 4 and 7 x 10^{13} cm.

SHOCK WAVE PROPAGATION

The outgoing shock wave that is thought to result from the collapse and bounce of the core (c.f. Wilson[10]) can then (if sufficiently strong) eject the mantle and envelope of the star, while the core of the star recollapses to form a neutron star or black hole. This shock also induces explosive nuclear burning in the region just above the core, which though crucial to the synthesis of the elements in stars, adds only about 10 to 20% to the energy of the final supernova explosion. Much of the silicon layer directly over the iron core reaches temperatures above 4 billion degrees and is burned into 0.1 to 0.4 $M_\odot$ of radioactive ^{56}Ni (the most strongly bound nucleus that can be formed from the nearly equal numbers of neutrons and protons present in the initial fuel). ^{56}Ni, however, beta decays first to ^{56}Co and then to ^{56}Fe with a half-life of 6.1 and 78.5 days, respectively. The γ-rays and positrons that result from these decays represent an important late-time input to the thermal energy of the exploding star.

As the shock wave continues to propagate out through the mantle ($\rho \gtrsim 10 \ g \ cm^{-1}$) of the star, the energy needed to heat and accelerate each successive mass shell is smoothly derived from the subsonic deceleration and adiabatic decompression of the underlying material. Sensitivity tests[6] show that except in the immediate vicinity of the mass cut between the mantle and the collapsed core, only the observable final energy of the outward-going shock wave and not the details of its formation are important in determining postshock conditions for a given presupernova configuration. The entropy of the postshock material is roughly constant due to the approximate cancellation of the centrally depressed (by neutrino emission) presupernova entropy profile by the centrally peaked profile of entropy production by the shock (due to shock deceleration). Since sufficient time is available for pressure balance to be established, this constant entropy profile results in nearly constant density and temperature profiles behind the shock front--in contrast to the thin, high density shells formed by a strong point explosion in an initially constant density medium.[11]

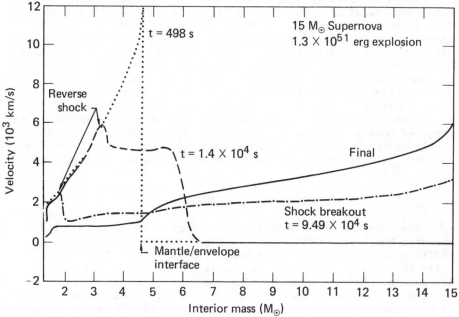

Fig. 2   Velocity as a function of interior mass coordinate in the 1.3 x 10^{51} erg, 15 $M_\odot$ supernova model at, and just after, the breakout of the shock wave from the surface. Each curve is labeled with the time in days since core collapse, and dotted line segments indicate regions with conditions allowing the growth of Rayleigh-Taylor instabilities.

Figure 2 details the hydrodynamical behavior of the mantle and envelope of our 15 $M_\odot$ model star undergoing a 1.3 x 10^{51} erg explosion (Model 15A, see Ref. 6). We shall discuss this case below as a typical Type II supernova because of its concordance with observation, and because 15 $M_\odot$ stars are observed to be more numerous than 25 $M_\odot$ stars[12], which we find to have generally similar behavior. During the time t=60-500 sec (measured from core collapse) the shock wave reaches and accelerates down the steep, 6-order-of-magnitude density gradient at the edge of the mantle, while the postshock material adiabatically expands and cools, initially unhindered by the surrounding low density envelope. By the time the exploding mantle has swept up sufficient envelope material to be significantly decelerated (t ~3,000 sec), its density has dropped to below 10^{-3} g cm^{-3} and its temperature below 2 x 10^6K. Under these conditions the expansion is hypersonic so that the deceleration of the mantle by the envelope is mediated by a reverse shock wave (see also Ref. 13).

This reverse shock has almost reached the center of the star when the principal shock wave reaches and accelerates down the final density gradient at the edge of the star at t=1.1 days. Subsequent adiabatic expansion over the next few days leads to the final velocity profile shown in Figure 2. Note that the velocity of the mantle has been reduced to below 1000 km/s both by the action of the reverse shock and the restraining pressure exerted by the decompressing envelope. The mantle, which contains virtually all the Z>2 elements synthesized by the star, including the [56]Ni, is thus left with only about 5% of total energy of the explosion.

Density and temperature profiles at, and just after, shock breakout are shown in Figures 3 and 4, and regions with conditions allowing the growth of Rayleigh-Taylor instabilities are indicated. In contrast to the more ad-hoc models studied by Chevalier and Klein[14], we find that most of the envelope is not subject to Rayleigh-Taylor instabilities during its acceleration. As pointed out by Lasher[15], however, the presence of such instabilities is likely to be a sensitive function of the density gradient in the convective red-giant envelope, and thus of our still incomplete understanding of superadiabatic convection.

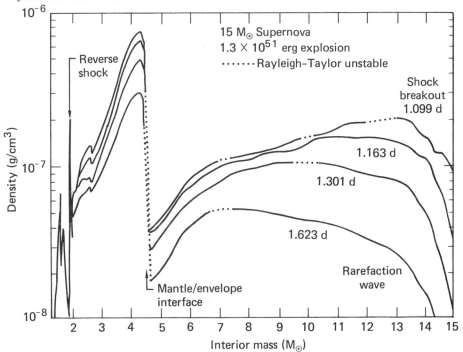

Fig. 3   Density as a function of interior mass coordinate in the 1.3 x 10[51] erg, 15 $M_\odot$ supernova model at, and just after, the breakout of the shock wave from the surface. Each curve is labeled with the time in days since core collapse, and dotted line segments indicate regions with conditions allowing the growth of Rayleigh-Taylor instabilities.

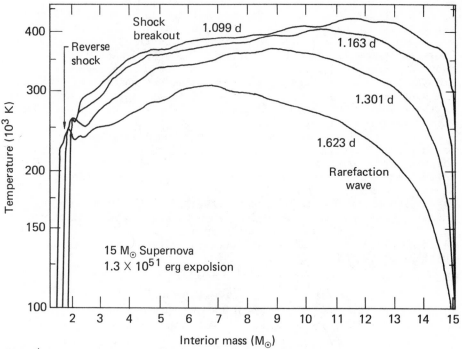

Fig. 4 Temperature profiles corresponding to the density profiles shown in Fig. 3.

TYPE II SUPERNOVA LIGHT CURVES

The light curves produced by two 15 $M_\odot$ supernova models (with differing explosion energies of 1.3 and 3.3 x 10^{51} ergs due to differing assumptions about detailed core physics) are shown in Figure 5, compared to photometric data for Supernova 1969ℓ in NGC 1058, perhaps the best observed Type II supernova.[2,16] SN 1969ℓ is characteristic of a common subclass of Type II supernovae which shows a 2-3 month initial plateau in its visual emission ($M_v \approx -17$) followed by a rapid decline of about 2 stellar magnitudes and then a slower decline at a rate of ~3 magnitudes/year. (SN 1970g in NGC 5457=M101 is another recent example of this subclass).[17] It is apparent that the theoretical and observational results are in excellent agreement. As Figure 4 shows, an increase in the energy of the core explosion produces a roughly linear increase in the optical brightness together with a shorter "plateau". The results for our 25 $M_\odot$ models are generally quite similar. Note in particular that this agreement has not been achieved by normalizing the observational absolute magnitude so as to provide the best fit.

Figures 6 and 7 compare derived observational SN 1969ℓ results [2,16,18] for the temperature and radius of the supernova photosphere with the corresponding theoretical results. Figure 8 shows a comparison between observed absorption line velocities and photospheric

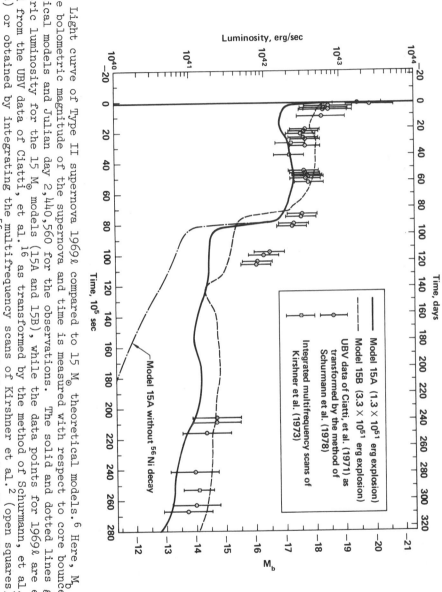

Fig. 5 Light curve of Type II supernova 1969ℓ compared to 15 M$_\odot$ theoretical models.[6] Here, M$_b$ is the absolute bolometric magnitude of the supernova and time is measured with respect to core bounce for the theoretical models and Julian day 2,440,560 for the observations. The solid and dotted lines give the bolometric luminosity for the 15 M$_\odot$ models (15A and 15B), while the data points for 1969ℓ are either derived from the UBV data of Ciatti, et al. 16 as transformed by the method of Schurmann, et al.[21] (open circles) or obtained by integrating the multifrequency scans of Kirshner et al.[2] (open squares). The dot—dashed curve shows the result when 56Ni decay is artificially suppressed in Model 15A.

22

velocities for the low energy 15 $M_\odot$ explosion. The agreement between theory and observation is uniformly within the observational errors, and allows a confident description of the general physical processes which are occurring (see also Ref. 13 and 19-22 for conclusions based on parameterized models).

The initial sharp spike in temperature and luminosity at small photospheric radius corresponds to the breakout of the supernova shock through the surface of the star's supergiant envelope. This is accompanied in theory[7,23-25] by a ~2000 second long, soft x-ray pulse with an equivalent black body temperature of ~1.5 x 10^5 K and a peak luminosity of ~10^{45} erg/second. The surface is then rapidly cooled by radiative emission and hydrodynamic expansion, balanced in part by radiative diffusion from below.

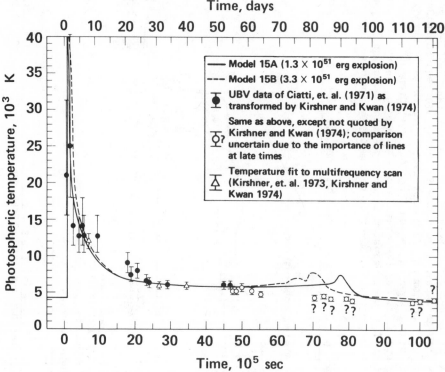

Fig. 6 Photospheric temperature of Type II supernova 1969ℓ compared with 15 $M_\odot$ theoretical models.[6] The solid and dotted lines give the results for the 15 $M_\odot$ models, while the solid circle data points represent the data of Ciatti, et al.[16] as transformed by Kirshner and Kwan.[18] The open circle data points are obtained by extrapolating the transformation methods of Kirshner and Kwan[18] to the late time data of Ciatti, et al.[16], and are associated with "?" marks where lines dominate the spectrum and such color/temperature transformations become dubious. The open triangle data points are derived from fits to multifrequency scans.[2]

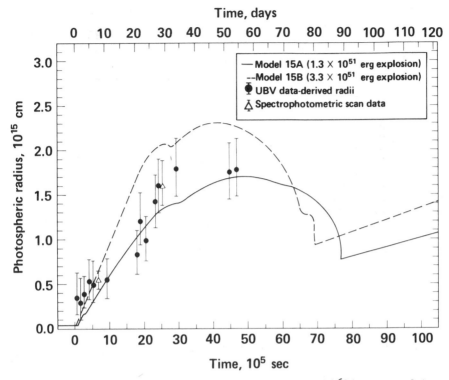

Fig. 7 Photospheric radius of Type II supernova 1969ℓ compared to 15 M⊙ theoretical models.[6] Solid and dotted lines give the results for the 15 M⊙ models, while the 1969ℓ data points are either derived from UBV data[18] or multifrequency scans.[2]

The star remains sufficiently optically thick during its acceleration that 99% of the total supernova energy is converted to kinetic energy of the expanding debris. Only about 1% thus remains to be radiated when the star finally starts to become optically thin after expanding to about 10^{15} cm, or about 10 to 30 times its initial radius. As Figure 5 illustrates, this division of energy is just sufficient to explain supernova observations of the SN 1969ℓ variety. Presupernova stars with radii much larger or much smaller than 3 to 7×10^{13} cm (corresponding to our 15 and 25 M⊙ models) would respectively radiate too great or too small a fraction of the total supernova energy to agree with the observations.

The photosphere initially lies close to the outer surface of the star, and thus expands rapidly in physical size as the star explodes. Figure 9 shows the evolution of the density profile near the surface of the star during the first 20 days of the explosion, and the position of the photosphere is noted. It is apparent that after the first few days the density at a given mass coordinate scales as t^{-3} corresponding to an homologous expansion (velocity ∝ radius) with a frozen-out velocity distribution. The photosphere first moves outward

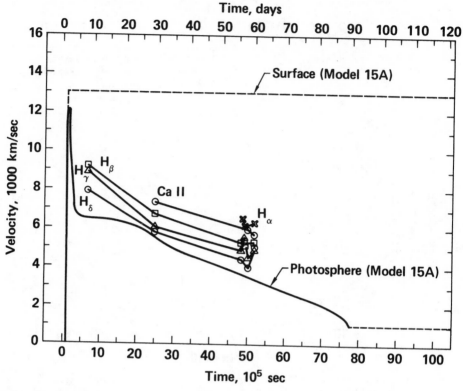

Fig. 8 Time evolution of the surface and photospheric velocities of the 1.3 x 10[51] erg, 15 $M_\odot$ supernova model (15A) compared with absorption lines velocities observed[2] for supernova 1969ℓ.[6]

in mass coordinate from its presupernova position due to shock-induced reionization of the overlying material, and then moves inward with respect to the expanding envelope material maintaining an almost constant density of 10^{-13} g/cc.

After about 20 days, the photosphere has cooled to roughly 6000 K, the temperature at which hydrogen plasma recombines to form a nearly transparent atomic gas at the extremely low densities prevalent in the envelope. This recombination-induced transparency allows the rapid radiative cooling of the immediately underlying layers, causing them in turn to recombine and become transparent. A cooling wave thus develops that sweeps inward through the exploding envelope over a period of about two months. The photosphere follows the recombination front associated with this cooling wave, and thus its temperature remains close to the 6000 K recombination temperature (see Fig. 6). In addition, the radius of the photosphere remains nearly constant ($\approx$1.5–2.0 x 10^{15} cm) during this period because the inward motion of the photosphere relative to the envelope material is approximately canceled out by the overall expansion of the

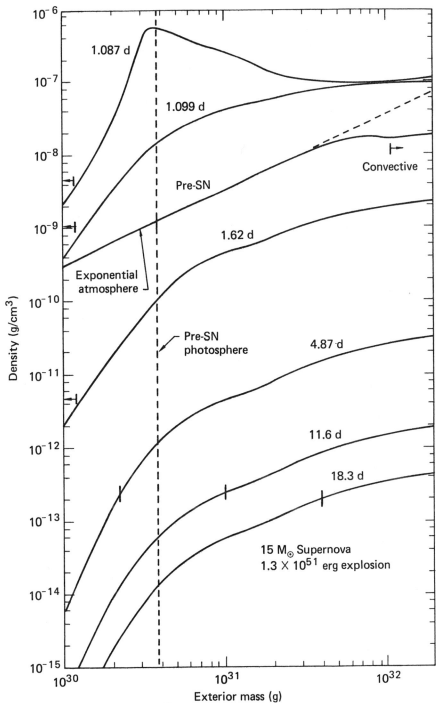

See following page for caption.

Fig. 9 (previous page) Density profiles as a function of exterior mass coordinate of the 1.3×10^{51} erg, 15 $M_\odot$ supernova model (15A) during the first 20 days of the explosion. The curves are labeled with the time since core collapse, and the position of the photosphere is indicated by a bar. The curve labeled "Pre-SN" is the presupernova density profile, which over most of the mass range shown has the form of an exponential atmosphere (dotted line). As indicated, the presupernova envelope is convective for exterior masses greater than 1.1×10^{32} g.

envelope. It is this combination of nearly constant photospheric radius and temperature that causes the supernova's luminosity to remain roughly constant during the plateau phase of the light curve (see Fig. 5).

Figures 10 and 11 illustrate the evolution of the temperature and density profiles during this epoch. Note that slight (possibly numerical) irregularities in the nearly homologous hypersonic flow have induced the growth of density spikes in the mantle.

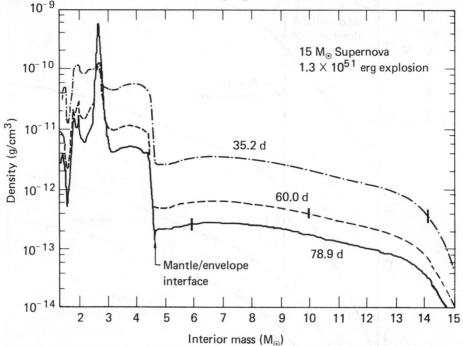

Fig. 10 Density as a function of interior mass coordinate in the 1.3×10^{51} erg, 15 $M_\odot$ supernova model during the transparency-wave-induced "plateau" epoch of its light curve. Each curve is labeled with the time since core collapse. The position of the photosphere is shown by the bar across each curve.

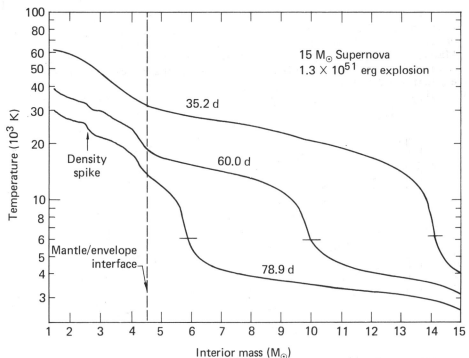

Fig. 11 Temperature profiles corresponding to the density profiles presented in Fig. 10. Again, the position of the photosphere is indicated by a bar.

Eventually, as the cooling wave encounters more slowly moving material deep within the envelope, the speed of the wave exceeds the matter velocity, and the photosphere physically shrinks until it encounters the slowly moving (<1000 km/s), relatively dense, and very optically thick mantle (see Fig. 7). At this point, a transient increase in the photospheric temperature occurs and persists for approximately two to three days as a result of the uncovering and rapid cooling of the hot surface of the mantle. This phenomenon is potentially observable, although the presence of strong emission lines in the overlying envelope may tend to mask it. The sharpness of this final recession and the resulting decrease in luminosity probably are artificially abrupt because of our relatively simplified treatment of recombination and emission-line effects.[5]

At late times, the luminosity results from the diffusion out of the mantle of thermalized radiation from the decay of the explosively generated ^{56}Ni and its daughter ^{56}Co. As Figure 5 shows, models in which ^{56}Ni decay is turned off display a much more sharply falling luminosity tail as the residual thermal energy in the mantle diffuses out over a characteristic time of only one to two months. In models containing energy output from radioactivity, temporary trapping of the thermalized decay energy in the optically thick mantle produces a luminosity decline slower than the 78-day ^{56}Co half-life, particularly in the 25 $M_\odot$ case.

The deposition of the positrons and γ-rays from these decays takes place almost entirely in the mantle due to the large column density of overlying material that is present even at very late times, as shown in Figure 12. As expected, the column density at a given mass coordinate scales as t^{-2} due to the nearly homologous expansion of the star.

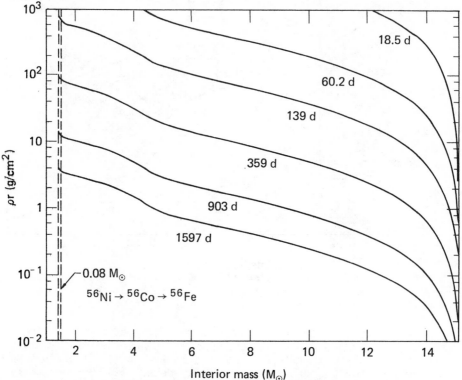

Fig. 12   Column density, ρr, of overlying material as a function of interior mass coordinate in the 1.3×10^{51} erg, 15 $M_\odot$ supernova model. Each curve is labeled with the time since core collapse, and the position of the 0.08 $M_\odot$ of ^{56}Ni formed in the explosion is indicated.

The peak emission of unscattered γ-rays takes place at about 400 days and yields a peak flux in the 3.2 MeV line of about 5×10^{-5} photons cm^{-2}sec^{-1}MeV^{-1} (see Ref. 26) for the 1.3×10^{51} erg, 15 $M_\odot$ supernova model, assuming a distance of 1 Megaparsec.

The ripples in the theoretical curves at late times result from the formation of density clumps in the mantle noted earlier, which if real appear to offer excellent sites for grain formation. It is likely that two-dimensional instabilities will also occur[14] (as is suggested by the clumpy appearance of supernova remnants) which should have the effect of increasing the noise, but damping the size of the transient excursions in the luminosity.

In principal, a great deal of information about supernova can be deduced from their light curves. At times earlier than 3 months, the light curve principally conveys information about the structure of the envelope of the presupernova star. More massive, extended envelopes produce prolonged luminosity plateaus and slower photospheric velocities for fixed shock energies. The behavior of the tail of the light curve, on the other hand, yields information about the size and density of the mantle, and ultimately, its composition.

In order to realize the full potential of these sources of information for understanding supernovae and the elements they produce, it will be necessary to perform detailed calculations of non-LTE supernova atmospheres so that spectral information about composition and plasma conditions can be unfolded. It is hoped that the Type II supernova models presented here will provide a starting point for, and help to motivate such work, as well as extensive new observations of supernovae. Taken together, such advances should allow the confident use of supernovae as both standard[27-28] and "non-standard candles"[18,21,29] for determining distances to distant galaxies.

EDGE-LIT CARBON DETONATIONS OF
ACCRETING WHITE DWARFS AS TYPE I SUPERNOVAE

Since this workshop is primarily concerned with Type I supernovae, it seems irresistable to broaden the topic of this talk to very briefly describe some of our recent calculations of the fate of white dwarfs undergoing slow mass accretion. A more detailed description will be published elsewhere.[30]

Our calculations were started from an initial model supplied by Taam[31] in which a 0.5 $M_\odot$ white dwarf, composed of equal concentrations of carbon and oxygen, accretes hydrogen (assumed burned quickly to helium) at a rate of 10^{-8} $M_\odot$ yr^{-1}, until 0.62 $M_\odot$ of helium has accumulated. The star's density and temperature profiles at this stage are shown in Figure 13. At this point, electron conduction cooling, which has allowed the helium to form a highly degenerate layer on the surface of the star ($\rho \sim 10^7$ g/cc, $T \sim 5 \times 10^7$ K) is no longer sufficient to counteract the compressional heating of the star due to the increasing overlying mass of helium, and a thermonuclear runaway results at the carbon/helium interface.

Using the stellar evolution/explosion code described above to follow this star's subsequent evolution, we found that, after a brief phase of convective helium burning, the overpressure from the nuclear runaway induces shock waves propagating both outward through the helium layer and inward through the carbon-oxygen core. These shocks ignite further nuclear burning and are rapidly transformed into self-sustaining Chapman-Jouguet detonation waves. The burning fuel is sufficiently inertially confined so that the final product is predominantly ^{56}Ni. The ease with which carbon is ignited in this manner, in contrast to the still unresolved difficulties[32,33] associated with the central ignition of carbon in white dwarfs near the Chandrasekhar mass, is a direct consequence of the 2-3 order-of-magnitude reduction in ignition density in the present case which makes possible nuclear-burning-induced overpressures of ~400%, instead of ~20-50%.

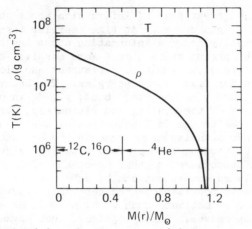

Fig. 13   Density (ρ) and temperature (T)   as   a   function   of
interior mass coordinate $(M(r)/M_\odot)$ in the initial white dwarf model
of Taam[31] just prior to He ignition.

The present status of this calculation is illustrated in Figure
14 which shows the diverging He and C/O nuclear detonation waves just
before they reach the surface and center of the star, respectively.
Preliminary extensions of this calculation suggest that the light
curve and velocities characteristic of Type I supernovae may result
as the ~ 1 $M_\odot$ of ^{56}Ni synthesized by the detonation, and expands and
decays.

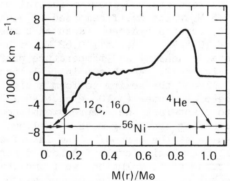

Fig. 14   Velocity and composition profiles of the detonating white
dwarf model.

A possible complication is that the He will most likely runaway
at a point instead of along an entire spherically symmetric mass
shell (as we are forced to assume), resulting in a detonation wave
propagating spherically outward from a point offset with respect to
the center of the star.   Generic two-dimensional investigations of
off-center detonations by Fryxell[34] suggest, however, that provided
the detonation can still progagate against the initially higher rate
of geometric divergence, the final state of the detonated star should
be nearly the same.

Ron Taam is gratefully acknowledged for supplying details of his accreting white dwarf models prior to publication and for useful discussions.

REFERENCES

1. R. Minkowski, Annual Reviews, 2, 247 (1964).
2. R. P. Kirshner, J. B. Oke, M. V. Penston, and L. Searle, Ap. J. 185, 303 (1973) and references cited therein; R. P. Kirshner, and J. B. Oke, Ap. J. 200, 574 (1975).
3. I. S. Shklovsky, Supernovae (Wiley, New York, 1968), p. 20.
4. J. Maza and S. van den Bergh 204, 519 (1976).
5. T. A. Weaver, G. B. Zimmerman, and S. E. Woosley, Ap. J. 225, 1021 (1978). Also see references therein.
6. T. A. Weaver and S. E. Woosley, Proc. 9th Texas Symposium on Relativistic Astrophysics, Munich, 1978, Ann. N.Y. Acad. Sci. 336, 335 (1980).
7. T. A. Weaver and S. E. Woosley (1980), in preparation.
8. S. E. Woosley and T. A. Weaver (1980), in preparation.
9. E. M. Burbidge, G. R. Burbidge, W. A. Fowler, and F. Hoyle, Rev. Mod. Phys. 29, 547 (1957).
10. J. R. Wilson and R. L. Bowers (1978), private communication; J. R. Wilson, Proc. 9th Texas Symposium on Relativistic Astrophysics, Munich, 1978, Ann. N.Y. Acad. Sci. 336, 358, (1980).
11. Ya. B. Zel'dovich and Yu. P. Raizer, Physics of Shock Waves and High-Temperature Hydrodynamic Phenomena (Academic, New York, 1966), p. 93.
12. G. E. Miller and J. M. Scalo, Ap. J. Suppl. 41, 513 (1979).
13. R. A. Chevalier, Ap. J. 207, 872 (1976).
14. R. A. Chevalier and R. I. Klein, Ap. J. 219, 931 (1978).
15. G. J. Lasher, Ap. J. 201, 194 (1975), and this conference.
16. F. L. Ciatti, L. Rosino, and F. Bertola, Mem. Soc. Astron. Ital. 42, 163 (1971).
17. R. Barbon, F. Ciatti, and L. Rosino, Astron. & Astrophys. 29, 57 (1973).
18. R. P. Kirshner and J. Kwan, Ap. J. 193, 27 (1974).
19. S. W. Falk and W. D. Arnett, Ap. J. Lett. 180, L65 (1973); S. W. Falk and W. D. Arnett, Ap. J. Suppl. 33, 515 (1976).
20. W. D. Arnett and S. W. Falk, Ap. J. 210, 733 (1976).
21. S. R. Schurmann, W. D. Arnett, and S. W. Falk, Ap. J. 230, 11 (1979).
22. E. K. Grassberg, V. S. Imshennick, and D. K. Nadyozhin, Astrophys. Space Sci. 10, 28 (1971).
23. G. J. Lasher and K. L. Chan, Ap. J. 230, 742 (1979).
24. R. I. Klein and R. A. Chevalier, Ap. J. Lett. 223, L109 (1978).
25. R. A. Chevalier and R. I. Klein, Ap. J. 234, 597 (1979).
26. T. S. Axelrod, private communication (1980).
27. L. Rosino, in Supernovae, ed. D. N. Schramm (Reidel, Dordrecht-Holland, 1977), p. 1.
28. L. Rosino and G. Di Tullio, in Supernovae and Supernova Remnants, ed. C. B. Cosmovici (Reidel, Dordrecht-Holland, 1974), p. 19.

32

29. D. Branch, in Supernovae, ed. D. N. Schramm (Reidel, Dordrecht-Holland, 1977), p. 21; R. V. Wagoner, Ap. J. Lett. 214, L5 (1977).
30. S. E. Woosley, T. A. Weaver, and R. E. Taam (1980), in preparation.
31. R. E. Taam, Ap. J. 237, 142 (1980) and references therein; and private communication (1979).
32. R. G. Couch and W. D. Arnett, Ap. J. 196, 791 (1975); S. W. Bruenn and A. Marroquin 195, 567 (1975) and references therein.
33. J. C. Wheeler, J-R. Buchler, and Z. K. Barkat, 184, 897 (1973); D. Sugimoto and K. Nomoto, preprint (1979), and references cited therein.
34. B. A. Fryxell, Ap. J. 234, 641 (1979).

TYPE I SUPERNOVAE: AN OBSERVER'S VIEW

Robert P. Kirshner
Department of Astronomy, University of Michigan

To an observational astronomer, a supernova is a new star of extremely high luminosity ($L > 10^8 L_\odot$). Generally these objects are discovered through supernova searches that use small Schmidt cameras to obtain repeated photographs of external galaxies, although they are sometimes discovered by chance as astronomers study other galaxies, and in historical times a handful have been sighted in our own galaxy (see Clark and Stephenson, 1978).

At the present, about 5 to 10 supernovae are discovered each year, and the total number since 1885 amounts to nearly 400 (Kowal, 1980). Each supernova is given a name that consists of the year of discovery, followed by a letter that indicates the order of discovery. Hence 1972e was the 5th supernova discovered in 1972. An important bibliographical source of information on individual supernovae is the compilation by Karpowicz and Rudnicki (1968). Two recent conferences on supernovae are also of general utility: see Cosmovici (1974) and Schramm (1977).

It is important to keep clearly in mind that the supernovae, in the sense of large optical outbursts, are not necessarily linked one-to-one with stellar deaths, pulsar formation, or the creation of black holes. Although a massive star is not likely to end its nuclear burning in a peaceful way, the optical outburst depends on the properties of the outside of the star. It is quite likely that some stars collapse to form compact stellar remnants, but release their energy at other wavelengths and are not counted as supernovae.

The classification of supernovae into types is a spectroscopic classification defined by a few prototypes in each class (Zwicky, 1968, Oke and Searle 1974). It is not defined by the shape of the light curve (luminosity vs. time) or by the type of galaxy in which the supernova erupted. For Type I, the prototypes are SN 1937c in IC 4182 (Minkowski, 1939, Greenstein and Minkowski, 1973) and SN 1972e in NGC 5253 (Kirshner et al. 1973 a, b, Kirshner and Oke, 1975.) In each of these cases, a supernova was discovered in a nearby galaxy and then observed frequently until it grew too faint for the best available detectors on the largest telescopes. In the case of 1972e, some spectroscopic data were obtained nearly two years after the explosion when the object was fainter than m = 21. An important difference between the 1937 data and the 1972 data is that the modern spectra are obtained photoelectrically rather than with photographic plates. This permits measurements to be made over a wide wavelength range from 3200 Å to 11000 Å and it also allows the data to be calibrated in absolute values of flux density: erg cm^{-2}s^{-1}Å^{-1}.

ISSN:0094-243X/80/630033-05$1.50 Copyright 1980 American Institute of Physics

In the 1972 data, the spectral resolution was no better than 20 Å and typically 40 or 80 Å. With current detectors, such as image dissectors, reticons, and vidicons it is possible to obtain the energy distribution of a supernova at 5-10 Å resolution over a span of about 3500 Å.

Some important qualitative results have emerged from the 1972 work that have had considerable impact on models for Type I supernovae. First, near maximum light most of the energy is carried in a continuum which looks very much like the energy distribution of a supergiant star. This continuum, which has an effective temperature near 12,000 K at maximum light (Kirshner, Arp, and Dunlap, 1976) cools to around 7000 K after a few weeks. During this epoch, the photosphere expands as the temperature drops. The expanding photosphere has a velocity near 10,000 km s^{-1}, a fact which indicates that supernovae really do explode, and which has been used to estimate the distance of extragalactic supernovae.

The distance estimate (Branch and Patchett, 1973, Kirshner and Kwan, 1974) comes from comparing the angular rate of expansion, as inferred from the photospheric temperature and the received flux measured at several epochs, with the linear rate of expansion as inferred from spectral lines. It is important to develop a correct picture of the expanding atmospheres of the supernovae in order to refine this distance estimate, since it is an extragalactic method that does not depend on the usual chain of methods leading from the distance to the Hyades through the calibration of the Cepheids and beyond.

With regard to the spectral lines observed near maximum light, a few appear to be the same lines that are present in SN II: Ca II, K and H, the Ca II infrared triplet, the Na I D lines (or pherhaps He I λ5876), and the Mg I b band. In SN II, the lines appear to form smaller excursion from the continuum than in SN I: this may result from SN I's having more line-forming ions compared to the source of continuum opacity. If the continuum opacity comes from electron scattering, this may just be a reflection of a low hydrogen abundance.

The most conspicuous difference between SN II and SN I lies in the strength of the hydrogen lines. In SN II, Hα and Hβ are clearly present, and very strong. In SN I, they may be entirely absent, and in any event are rather weak.

Most of the strong lines in SN atmospheres are present both in absorption and in emission. Generally, a violet-shifted absorption trough extends out to blue shifts of ~15,000 km s^{-1} or more, while an emission peak is generally observed near and to the red of the rest wavelength of a line. This is the type of line profile observed in ordinary stars that have substantial mass outflow: they are called P Cygni lines after the prototype star. Physically, P Cygni lines arise in an expanding atmospher that scatters, but does not absorb, photospheric photons.

The problem of line identifications in SN I near maximum light is made complex by overlapping emissions and absorptions. Models by Branch overcome some of this confusion by synthesizing large regions of the spectrum rather than concentrating on individual features. Doing this correctly requires large amounts of atomic data on low-lying levels of low ionization stages for common elements.

At late times, after the continuum has faded into insignificance, the spectra of SN I are dominated by four broad bands of emission at $\lambda4200$, 4600, 5000, and 5300. The famous exponential decay in the light curve of SN I in blue light is an observation that refers to the behavior of this complex. In 1975, Kirshner and Oke suggested that [Fe II] lines, lots of them, might blend together to form some of these emission bands. Refinements of that idea are presented in this conference by Meyerott and by Axelrod.

The really tempting idea that SN I synthesize iron-peak nuclei that provide energy through radioactive decay and can be observed in the spectrum at late times requires more stringent observational tests. As a small contribution to the discussion, I have compiled a table which gives a good estimate of the energy emitted by an SN I as a function of time. Near maximum light ($t \leq 46$ days), I have integrated under the blackbody that best fits the spectrum in the range 3300 - 11,000Å. Since this wavelength range includes the peak of the blackbody curve, and there is no clear indication of large fluxes in the ultraviolet (Holm, Wu, and Caldwell, 1974) or infrared (Kirshner et al. 1973b), this seems reasonable.

At later times, a blackbody is a very poor representation of the spectrum. Since there is no sensible way to include flux that is outside the observed wavelength region, I have merely summed the received energy from 3000 - 11,000 Å. Each data point should be good to about 20%.

I have used distances of 45 Mpc for NGC 2207 (SN 1975a; Kirshner, Arp, and Dunlap 1976) and 4 Mpc for NGC 5253 (SN 1972e). If these distances are incorrect, then the emitted power is wrong by the square of the distance.

TABLE 1

SN I LUMINOSITY

days	Age	10^6 sec	log L (erg s^{-1})
		SN 1975a in NGC 2207	
- 5		-0.43	43.23
- 2		-0.17	43.28
		SN 1972e in NGC 5253	
+ 14		1.21	42.94
+ 25		2.16	42.76
+ 36		3.11	42.32
+ 46		3.97	42.23
+ 82		7.1	41.9
+206		17.8	40.8
+237		20.5	40.5
+349		30.2	39.9
+418		36.1	39.5
+716		61.9	38.3

Finally, let me make a suggestion to those who search for supernovae. I believe that the discussion at this conference demonstrates the importance of finding bright supernovae that can be followed intensively for long stretches of time. Obtaining ultraviolet, infrared, and better optical data on a new bright SN I will add more to the discussion than 100 new SN at the limit of spectroscopic study.

RPK's research on supernovae is supported by the National Science Foundation through grant AST 77-17600 and by an Alfred P. Sloan Foundation Research Fellowship.

REFERENCES

Branch, D., and Patchett, B. 1973, M.N.R.A.S., 161, 71.

Clark, D. H., and Stephenson, F. R. 1977, The Historical Supernovae (Oxford: Pergamon).

Cosmovici, B. C. (ed) 1974, Supernovae and Supernova Remnants (Boston: Reidel).

Greenstein, J. L., and Minkowski, R. 1973, Ap. J., 182, 225.

Holm, A. V., Wu, C. C., and Caldwell, J. 1974, Pub. A.S.P., 86, 296.

Karpowicz, M., and Rudnicki, K. 1968, Preliminary Catalog of Supernovae (Warsaw: Warsaw Univ. Press).

Kirshner, R. P., Oke, J. B., Penston, M., and Searle, L. 1973a, Ap. J., 185, 303.

Kirshner, R. P., Willner, S. P., Becklin, E. E., Neugebauer, G., and Oke, J. B. 1973b, Ap. J. (Letters), 180, L97.

Kirshner, R. P., and Oke, J. B. 1975, Ap. J., 200, 574.

Kirshner, R. P., Arp, H. C., and Dunlap, J. R., 1976, Ap. J., 207, 44.

Kowal, C. T. 1980, Master List of Supernovae, Caltech.

Minkowski, R. 1939, Ap. J., 89, 143.

Oke, J. B., and Searle, L. 1974, Ann. Rev. Astron. Astrophys., 12, 315.

Schramm, D. N. (ed.) 1977, Supernovae (Boston: Reidel).

Zwicky, F. 1968, in Stars and Stellar Systems, Vol. VIII (Chicago: Univ. of Chicago Press).

SYNTHETIC SPECTRA OF SUPERNOVAE

David Branch*
University of Texas, Austin, TX 78712

ABSTRACT

The predicted characteristics of individual line profiles and blends formed by resonance scattering in supernova envelopes are described. Synthetic spectra consisting of blended scattering profiles on an underlying continuum are compared with McDonald Observatory spectra of a Type I and a Type II supernova. The synthetic spectra account reasonably well for many of the features observed during the early phases of both types, although not for the Balmer lines in the Type II. Some rough constraints on the density profiles and chemical compositions of the observable layers of the envelopes are given.

INTRODUCTION

Progress toward a more quantitative understanding of the Doppler-broadened, blended spectra of supernovae will require the comparison of observed and synthetic spectra. Synthetic spectra calculated until now have consisted of a continuum with super-imposed emission lines[1], absorption lines[2,3], or both[4], and emission lines without a continuum[5,7]. Although pure emission lines may account for at least the strongest features in late-time spectra of Type I SN[5-7], synthetic spectra consisting of P Cygni-type lines (emission with blueshifted absorption) are needed to represent Type II and the early phases of Type I.

I have begun a program to compute such spectra and to compare them with McDonald Observatory spectra of real SN. The resolution and signal-to-noise ratio of the McDonald spectra are sufficient for detailed studies of the lines. The calculations which have been carried out so far are based on a very simple model of spectrum formation which nevertheless is apparently adequate to account fairly well for many of the observed features and to provide some useful constraints on the physical conditions in SN envelopes.

THE MODEL

Spectral lines are assumed to be formed by scattering in a spherically symmetric, homologously expanding envelope above but near a well-defined photosphere which emits a blackbody continuum. Line formation is treated in the escape-probability approximation[8,9], which simplifies the radiative transfer by recognizing that when an emitted line photon escapes from its region of origin

*On leave from University of Oklahoma, Norman, OK 73019

in a rapidly expanding envelope, it Doppler shifts out of resonance with its parent transition and escapes completely unless it
strikes the photosphere and is absorbed there. A line profile can
be calculated when the radial dependence of the expansion velocity,
the line optical depth, and the line source function are specified.
The velocity is assumed to increase in proportion to the radius,
as expected after an explosion. The line optical depth is given
approximately by

$$\tau(r) = \frac{\pi e^2}{mc} f \lambda N(r) t \tag{1}$$

where N is the population of the lower level, t is the time since
outburst, and the other symbols have their usual meanings. Rough
estimates indicate that the ionization balance in the envelope
should be controlled by radiative processes. The ionization
balance should not change rapidly with radius owing to the op-
posing effects of the falling electron density and the decreasing
geometrical dilution factor for ionizing radiation from the photosphere. Thus the line optical depths should be controlled mainly
by the density profile, and all optical depths of interest should
decrease outward. For simplicity, all line optical depths are
taken to vary as r^{-n}. The source function is taken to be that of
resonance scattering,

$$S(r) = W(r) B_\lambda(T) \tag{2}$$

where $W(r)$ is the dilution factor and $B_\lambda(T)$ is the Planck function
for the temperature at the photosphere. Feldt[10] has found that
the predicted source function, based on solutions to the statistical rate equations for hydrogen in Type II envelopes and Si II
in Type I, does not differ radically from the resonance scattering
case.

Line profiles calculated for three values of n are shown at
the top of Fig. 1. Synthetic spectra calculated with n=4 turn out
to be unsatisfactory because lines of moderate strength are predicted to show extended blue absorption wings which are not observed. Spectra with n=7 and n=12 fit about equally well. As a
working value n=7 has been adopted, simply because an analytic
hydrodynamical result[11] gives n=7 if the density is forced to
follow a power-law. The profiles at the bottom of Fig. 1 show
that for n=7 the blueshift of the absorption minimum corresponds
closely to the velocity of the matter at the photosphere unless
the line is very strong.

The proper calculation of a P Cygni blend is a complicated
problem of multiple scattering because a photon which escapes from
its region of origin can redshift into resonance with and be scattered by other transitions of longer wavelength. Fortunately
Castor and Lamers[12] have shown that a simple approximation for

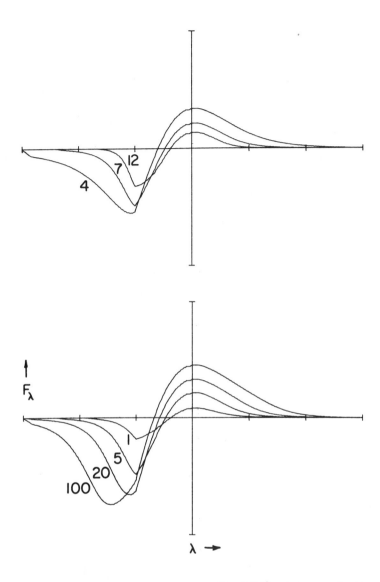

Fig. 1. Top: line profiles for $\tau = 5(r/R)^{-n}$ with n = 4, 7, and 12. Bottom: for $\tau = \tau_R(r/R)^{-7}$ with τ_R = 1, 5, 20, and 100. The unit on the horizontal axis is the wavelength shift corresponding to the velocity at the photosphere.

calculating a blend works quite well. Several minor modifications to their blend prescription have been made, to allow for the fact that in SN the lines are formed near the photosphere whereas Castor and Lamers were mainly concerned with line formation farther from the photosphere in extended stellar atmospheres. Fig. 2 shows a calculated blend of two lines having identical optical depth laws but separated in rest wavelength by an amount which corresponds to the velocity at the photosphere. Note that the blend has a peak at a wavelength which does not correspond to the rest wavelength of either line, while both absorption minima do occur at the expected blueshifted wavelengths. This shows clearly why it is better to use absorption minima rather than emission peaks for establishing line identifications by wavelength coincidence. Of course when synthetic spectra are matched to the observations all the information can be used.

A spectrum calculation consists of choosing the temperature of the underlying blackbody continuum, then blending line profiles in order of increasing wavelength so that each line operates on all lines of shorter wavelength, and finally applying interstellar reddening and instrumental broadening when required. For each ion of interest the optical depth at the photosphere of the strongest line is a fitting parameter, and the relative optical depths of

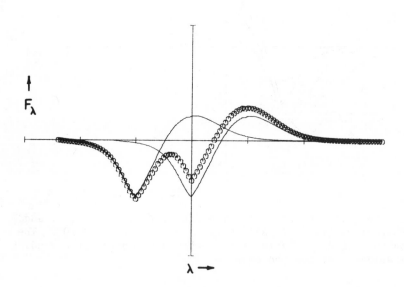

Fig. 2. A blend of two lines.

the other lines are fixed by assuming Boltzmann excitation at the temperature at the photosphere.

A SPECTRUM OF A TYPE II SUPERNOVA

A spectrum of SN 1979c in M100, observed about 45 days after maximum light, and a synthetic spectrum are shown in Fig. 3. The observed spectrum contains terrestrial absorption near 6800 Å and unresolved interstellar absorption near 5900 Å; the other features are those of a normal Type II except that the Hα profile is unusual in lacking an obvious blueshifted absorption. Clearly the hydrogen lines, especially Hα, are not adequately represented by resonance scattering profiles, but the feature identified as the Na I D lines does resemble a resonance scattering profile. All of the other features in the synthetic spectrum are produced by blends of Fe II lines. Despite the poor representation of the Balmer lines, the wavelengths of the absorptions in the synthetic spectrum all match those of the observed spectrum rather well. This, as well as the near-blackbody energy distribution[13], supports the assumption that the photosphere is well defined. The identification of the remaining absorption, near 5400 Å, must be sought by introducing another ion (Sc II?)

Similar comparisons for a series of spectra of SN 1979c show that from one to six weeks after maximum light the temperature at the photosphere fell from 11,000 to 5500°K and the velocity of matter at the photosphere decreased from 10,000 to 8000 km/sec as the photosphere receded with respect to the expanding material. The temperatures and velocities are useful for comparison with hydrodynamical models [14-16] and for making an independent determination of the distance to the supernova[13] . The line optical depths, together with estimates of the excitation and ionization conditions at the photosphere, imply roughly solar abundances, as expected for the outer layers of a massive star.

The outstanding problem with the optical spectrum of SN 1979c is to understand the Hα profile. Level populations calculated by solving the rate equations do not yet explain why the source function is so different from that of resonance scattering[10,17].

SPECTRA OF A TYPE I SUPERNOVA

Observed spectra of SN 1972e in NGC 5253[18] are compared with synthetic spectra in Fig. 4. The vertical marks above and beneath the spectra are for a discussion of line identifications by wavelength coincidence and can be ignored here. Terrestrial absorption occurs near 6800, 7200, and 7600 Å, and unresolved interstellar absorption near 3950 and 5900 Å. The May 18 spectrum refers to one or two weeks after maximum light, during the initial rapid decline of the light curve. Features in the synthetic

44

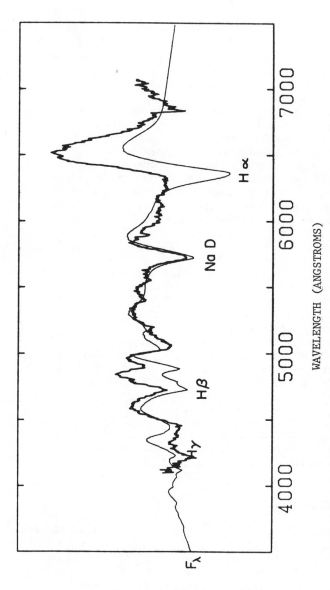

WAVELENGTH (ANGSTROMS)

Fig. 3. Observed spectra: the Type I SN 1972e in NGC 5253, obtained at 8Å resolution by R. G. Tull
Derek Wills using the McDonald Observatory 2.7-m telescope and the intensified-dissector-scanner
spectrograph constructed by Paul Rybski. Synthetic spectrum: V = 8000 km/sec, T = 5500 °K, lines of
hydrogen, Na I, and Fe II.

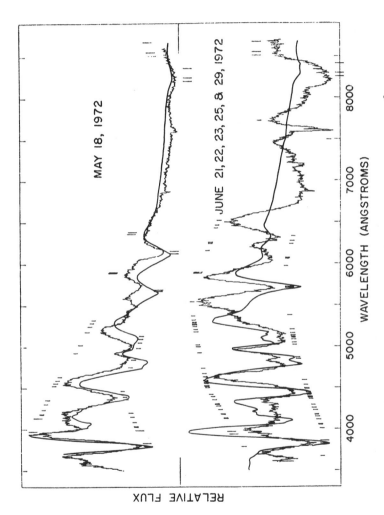

Fig. 4. Observed spectrum: the Type I SN 1972e in NGC 5253, obtained at 8Å resolution by R. G. Tull using the McDonald Observatory 2.7-m telescope and the coude scanner. Upper synthetic spectrum: V = 10,900 km/sec, T = 15,000 °K, A_V = 0.3, lines of He I, Ca II, Si II, and Fe II. Lower synthetic spectrum: V = 8000 km/sec, T = 6000 °K, A_V = 0.3, lines of Na I, Ca II, and Fe II.

spectrum are produced by Ca II (the H and K blend near 3950 and
the triplet near 8600 Å), Si II (doublets at 4130 and 6335 Å),
He I (5876 Å), and blends of many Fe II lines. The absorption
feature near 5700 Å is not at the predicted wavelength for either
the He I line or the Na D lines; a possible explanation is that
partial thermalization of the population of the highly excited
lower level of the He I line causes the optical depth to decrease
very rapidly above the photosphere. Another ion (S II?) will
need to be introduced to account for the absorption near 5500 Å;
it remains to be seen whether another ion can also remove the discrepancy near 5200 Å.

The June spectrum refers to 40-50 days after maximum light,
just after the end of the initial decline of the light curve. At
this phase the presence of a photosphere becomes questionable,
but the synthetic spectrum does account for the positions of some
of the features. In particular, note that a strong 4600 Å peak
can be produced by blends of Fe II permitted lines formed by scattering. In the two synthetic spectra, the changes in the Fe II
blends near 5000 Å are produced by increasing the Fe II optical
depths by a factor of two and reducing the velocity from 10,900
to 8000 km/sec. The feature at 6500 Å may be due either to Fe II
or Hα; conflicting values of the oscillator strengths of the relevant Fe II lines have been published recently[19], [20]. The infrared Ca II triplet is computed too weak relative to H and K. A
detailed study of Ca II level populations in SN envelopes would
be valuable.

Kirshner and Oke[5] and Meyerott[7] have shown that the strongest
features in the late-time spectra of SN 1972e can be interpreted
in terms of [Fe II] and [Fe III] emission line blends. It is very
likely that P Cygni scattering features also play a role. The two
absorption features attributed to Ca II and the one due to He I
and/or Na I persist in the spectrum throughout the first several
hundred days[21], [5]. The profiles of these features in the spectrum
obtained by Kirshner and Oke about 250 days after maximum light
can be accounted for nicely by resonance scattering in a shell
having $8000 \leq v \leq 18,000$ km/sec, surrounding a small continuum
source. It is interesting that scattering by permitted Fe II
lines provides a good fit to the positions of many features in the
250-day spectrum; however the calculated intensities in the blue
part of the spectrum are too low relative to the rest of the spectrum. Unless the scattering occurs on a continuum which is
strongly peaked in the blue, these Fe II coincidences may be
accidental.

During much of the initial peak of the light curve the velocity of matter at the photosphere appears to remain constant near
11,000 km/sec. Absorption in the Ca II lines extends out to
about 18,000 km/sec. At later phases we see evidence of material

moving as slow as 8000 km/sec. The picture which emerges is that of an ejected shell of material with appreciable density over the range 8000-18,000 km/sec and a density peak at 11,000 km/sec. The scattering profiles provide no direct evidence for matter moving slower than 8000 km/sec.

The optical depths needed for the synthetic spectra suggest a helium-rich composition, but with the mass fractions of the observable heavier elements somewhat enhanced relative to solar abundances.*

DISCUSSION

An understanding of supernova spectra leads to estimates of the density and chemical composition of the ejected matter as functions of velocity, which are badly needed for attempts to identify the SN progenitors. To improve these estimates the effects of including additional ions in the spectrum calculations must be explored, and some of the simplifying assumptions must be removed. One of the first steps will be to drop the assumption that the optical depth law is the same for all lines by including estimates of the radial changes in excitation and ionization. Another step will be to examine more closely the consequences of the artificial distinction between continuum- and line-forming layers.

For the spectrum calculations described here, the only atomic data required are oscillator strengths for strong permitted lines, especially for singly ionized elements of the iron group. When the physical conditions in SN envelopes become more precisely known it will be worthwhile to solve the rate equations for all level populations of interest; the limiting accuracy will be determined by our knowledge of photoionization cross sections from excited states[10].

This work was supported by NSF Grant AST-7808672.

REFERENCES

1. F. L. Whipple and C. Payne-Gaposchkin, Proc. Am. Phil. Soc. 84, 1 (1941).
2. B. Patchett and D. Branch, Mon. Not. Roy. Astr. Soc. 158, 375 (1972).

*After the Workshop evidence for the presence of Co II lines was sought in the May and June spectra, using estimates of the oscillator strengths of the strongest Co II lines in the visible spectrum[22,23]. These lines do not contribute significantly to the SN spectra; the cobalt mass fraction in the material moving faster than 8000 km/sec is probably less than 0.01.

3. D. Branch and B. Patchett, Mon. Not. Roy. Astr. Soc. 161, 71 (1973).
4. D. Branch and J. L. Greenstein, Astrophys. J. 167, 89 (1971).
5. R. P. Kirshner and J. B. Oke, Astrophys. J. 200, 574 (1975).
6. G. E. Assousa, C. J. Peterson, V. C. Rubin, and W. K. Ford, Jr., Publ. Astr. Soc. Pac. 88, 828 (1976).
7. R. E. Meyerott, Astrophys. J., in press.
8. V. V. Sobolev, Moving Envelopes of Stars (Harvard University Press, 1960).
9. J. I. Castor, Mon. Not. Roy. Astr. Soc. 149, 111 (1970).
10. A. N. Feldt, Ph.D. Thesis, University of Oklahoma (1980).
11. S. A. Colgate and C. McKee, Astrophys. J. 157, 623 (1969).
12. J. I. Castor and H. G. J. L. Lamers, Astrophys. J. Suppl. 39, 481 (1979).
13. D. Branch, M. McCall, P. Rybski, A. Uomoto, B. Wills, and D. Wills, Bull. Am. Astr. Soc. 11, 694 (1979).
14. R. A. Chevalier, Astrophys. J. 207, 872 (1976).
15. S. W. Falk and W. D. Arnett, Astrophys. J. Suppl. 33, 515 (1977).
16. T. A. Weaver and S. E. Woosley, Paper presented at the 9th Texas Symposium on Relativistic Astrophysics, Munich (1978).
17. R. I. Klein and R. A. Chevalier, Private communication.
18. D. Branch and R. G. Tull, Astron. J. 84, 1837 (1979).
19. M. M. Phillips, Astrophys. J. Suppl. 39, 377 (1979).
20. J. B. Oke and T. R. Lauer, Astrophys. J. 230, 360 (1979).
21. R. P. Kirshner, J. B. Oke, M. V. Penston, and L. Searle, Astrophys. J. 185, 303 (1973).
22. R. A. Roig and M. H. Miller, J. Opt. Soc. Am. 64, 1479 (1974).
23. R. L. Kurucz and E. Peytremann, Smithsonian Astrophys. Obs. Spec. Rep. No. 362 (1975).

THE EXCITATION OF SPECTRA IN THE ENVELOPES OF SUPERNOVAE
AT LATE TIMES BY THE DEPOSITION OF POSITRONS AND γ-RAYS

Roland E. Meyerott
La Jolla Institute
La Jolla, California 92038

ABSTRACT

The excitation of spectra in the envelopes of supernovae at late times by the deposition of positrons and γ-rays is reviewed. It is shown that the species concentrations are determined by the high energy positrons or Compton electrons. Most of the energy of the secondary electrons created in the ionization process by the high energy primary electrons is deposited in the electron gas. The temperature of the electron gas is determined by the rate of energy deposition by the secondary electrons and the rate of cooling by radiation. In Type I Supernovae envelopes at late times, the principal source of cooling is forbidden line radiation. In Type II Supernovae envelopes, charge and energy transfer collisions with H^+ and He^+ may be important in determining the excitation and ionization states of the minor species. The outstanding problems in atomic plysics and spectroscopy that limit the application of this model to the excitation of spectra in supernovae envelopes are indicated.

I. INTRODUCTION

It is becoming increasingly apparent that some additional source of energy is required to maintain the late time luminosity of supernovae (SN). The most likely candidate for this late time energy source is the energy from the explosively synthesized ^{56}Ni, which decays as ^{56}Ni $\xrightarrow{6.1d}$ ^{56}Co $\xrightarrow{77d}$ ^{56}Fe [1].

In the previous work [1,2,3], radioactivity decay energy was added to the internal energy of the plasma and the resulting luminosity vs. time was computed. Little or no account was taken of non-LTE effects. The neglect of non-LTE effects in computing the light curve is not likely to be serious, since all of the energy deposited in the SN envelope at late times must be radiated (perhaps with some time delay) or else the envelope becomes unreasonably hot. The late time light curve is determined by the fraction of the radioactive energy deposited in the SN envelope and the fraction of the total luminosity that is radiated in the observable spectral regions [2,3,4].

Non-LTE effects and the details of the deposition of radioactive energy are important in the determination of the spectra emitted at late times. The spectral data indicate that the electron temperatures at late times are ~ 5000° K [5,6,7]. However, an electron of temperature of ~ 5000° K is too low to produce all the observed species concentrations and excitations by electron collisional mechanisms. Ionization and excitations by MeV electrons from the radio-

ISSN:0094-243X/80/630049-26$1.50 Copyright 1980 American Institute of Physics

active energy deposit are also important in determining the late time
spectra. In this paper the excitation of the spectra in the envelopes of SN's by the deposition of positrons and γ-rays will be analyzed
The outstanding problems in atomic physics and spectroscopy are
indicated.

II. MECHANISMS FOR THE EXCITATION
OF LATE TIME SPECTRA OF TYPE I SUPERNOVAE

The similarity of the late time spectra of Type I supernovae
(SN I) to the synthetic spectra of Fe^+, calculated assuming a
Boltzmann population of the metastable levels at a temperature of ~
5000° K, has been pointed out by Kirshner and Oke[5]. For the SN
1972e in NGC 5253 spectral region 4200-5500 A, which contains most of
the emission for times longer than 200 days, they obtained good
qualitative agreement between the predicted Fe II synthetic spectrum
and the observed features at 4300 A, 4800 A, and 5200 A. Only the
feature at 4600 A, the most intense feature of the spectrum, is not
predicted.

The feature at 4600 A has been identified by Gordon[8] as due to
Fe III. Gordon suggests that the electron temperature would be
8000° K at times 365 d after the explosion. This feature has also
been identified by Chiu, Morrison, and Sartori[9] as the $P\alpha$ line of
He II. They state that this identification is basic to the fluores-
cence-reverberation model[10] for the late time spectra of SN I.
Abdulwahab and Morrison[10] have noted that Fe^{++} has a multiplet that
coincides with the 4600 A feature in SN I spectra. They also note
the qualitative agreement between the spectra of SN 1972e and a combined synthetic spectrum of Fe^+ and Fe^{++}, assuming a Boltzmann population of the metastable levels at a temperature of ~ 1 eV. They
rule out the possible identification of the 4600 A feature in SN I
spectra as the 4600 A multiplet of Fe^{++} because of the small continuous red shift exhibited by the 4600 A feature.

Meyerott[7] has calculated the synthetic spectra of Fe^+ and Fe^{++}
as a function of electron temperature, assuming a Boltzman population of metastable levels. These synthetic spectra are compared to
the spectra of SN 1972e in NGC5253 taken 245 d after maximum[11]. It
is shown that good quantitative agreement can be obtained between
the observed spectrum and the synthetic spectrum at an electron temperature of ~ 4000° K with a concentration ratio of Fe^{++} to Fe^+ of
~ 4 to 1. In Figure 1a is shown the ratio of the observed flux at
255 d after the explosion radiated in 40 A to the total radiated
from 3220 A to 11,600 A plotted as a function of wavelength. In
Figure 1b is shown a synthetic spectra calculated from the synthetic
spectra of Fe^+ and Fe^{++} at an electron temperature, T_e = 4000° K
with a concentration ratio of 4 Fe^{++} to 1 Fe^+. This spectra has
been normalized so that the total integral from 3320 A to 11,600 A
is 0.6 rather than 1 for the observation. The other 0.4 of the flux
must be contributed by processes other than the forbidden lines,
such as radiative recombination. It can be seen that there is good
quantitative agreement between the principal features of the synthetic spectrum and the observed spectrum.

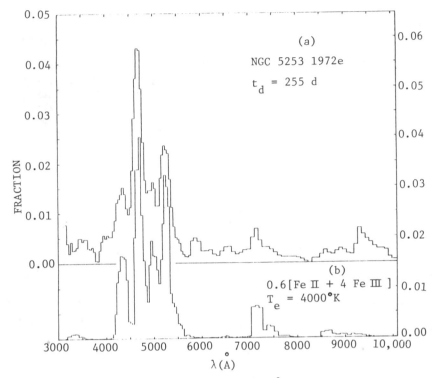

Fig. 1a Ratio of the flux radiated in 40 Å to the total radiated
from 3220 Å to 11,600 Å versus wavelength at t_d = 255 d.[7]

Fig. 1b Synthetic spectrum of Fe II + 4Fe III at T_e = 4000° K,
normalized, so that the total integral from 3220 Å to
11,600 Å equals 0.6.

The presence of Fe^{++} with a concentration ratio of 4 times that
of Fe^+ indicates that the ionic species concentrations are not determined by electron collisions from the thermalized electrons. The
ionization potential of Fe^+ is 16.16 eV, so that the collisional excitation rate is much too small to produce the observed Fe^{++} concentrations. The ionization must be produced by high energy electrons
which arise from the radioactive source.

The ionizing radiations from the radioactive decay of ^{56}Ni and
^{56}Co are γ- and positron radiation of energy ~1 MeV. When the γ-rays
Compton scatter they produce electrons with ~1/2 the γ energy.
Hence, one can consider the ionizing source to be MeV electrons
interacting with the gas of the SN envelope. The energy loss of
high energy electrons is not a strong function of the nature of the[7]
atoms or their stage of ionization. It has been argued by Meyerott[7]
that the fact that no H or He is observed in the late time spectra
of SN I's indicates that the principal energy deposition must take
place in the core materials, Fe, Co, Si, S, Mg, and Ca. This is not
an unreasonable suggestion in view of the fact that the radioactive

atoms are produced in the core materials during the explosive synthesis process.

It will be assumed that all the radioactive energy that is deposited in the expanding envelope of the SN is deposited in the core materials consisting of the heavy metals SI, Mg, S, Ca, Fe, and Co. Since the ionization potentials of the above species are all approximately the same (at least for the first few stages of ionization), these metals will be lumped together and treated as a single species. The ionization potentials are taken to be those of Fe and recombination coefficients taken to be the same as the Mg ions, since those for the Fe ions are not available.

TABLE I

LUMINOSITY OF 1972e AS A FUNCTION OF TIME

t_d (d)	$\mathcal{L}(t_d)$ (erg s^{-1})	$\mathcal{L}(t_d)$ (eV s^{-1})
78	8.99×10^{41}	5.62×10^{53}
255	4.01×10^{40}	$2.51 + 52$
435	4.20×10^{39}	$2.63 + 51$
735	(7.87×10^{37})	$4.92 + 49$

The energy deposited in the SN envelope per sec will be taken to be equal to the luminosity of 1972e. In Table 1 is shown the luminosity of 1972e in the wavelength band from 3320 Å to 10,600 Å. The data are from Oke[11]. The time, t_d, in the table, is the time in days after the explosion. The time to maximum, t_m, is taken to be 10 d, hence $t_d = t_m + 10$. The distance to the SN was taken to be 3.5 Mpc, as determined by Sersic, Carranza, and Pastoriza[12]. No corrections for interstellar absorption have been applied. The value at 725 d is extrapolated from the other measurements. Since little energy was observed in the ultraviolet[13] or in the infrared[14] after 68 d, these values probably represent the total energy radiated by the SN. The total energy deposited is such that, if it is not radiated, the SN would become extremely hot[15].

The SN envelope is assumed to be a uniform density shell of gas expanding with a mean velocity, u x 10^8 cm s^{-1}, with a velocity spread of $\pm$ 0.1u x 10^8 cm s^{-1}. This assumed gas distribution is consistent with that calculated by Lasher[16], using a hydrodynamics code. The volume of the shell at time t_d days after the explosion is

$$V = 1.63 \times 10^{39} u^3 t_d^{3}.$$

(1)

TABLE II DEPENDENCE OF VARIOUS QUANTITIES DEFINED IN TEXT ON u

u	3	5	7	10
$V(255)$ cm^3	7.30 + 47	3.38 + 48	9.24 + 48	2.70 + 49
$\mathcal{E}_d(255)$ eV cm^{-3} s^{-1}	3.34 + 4	7.44 + 3	2.70 + 3	9.28 + 2
$N(255)$ cm^{-3}	5.04 + 6	2.35 + 6	1.41 + 6	8.29 + 5
$\eta(255)$	3.68 + 54	7.94 + 54	1.30 + 55	2.24 + 55
$\eta_{Fe}(255)$	3.98 + 55	5.91 + 55	9.03 + 55	1.42 + 56
$M/M_0 \; \xi = 1$	1.71 − 1	3.70 − 1	6.09 − 1	1.04
$M/M_0 \; \xi = 2$	1.28 − 1	2.78 − 1	4.55 − 1	7.80 − 1
$M_{Ni}^s/M_0 \quad \xi = 1$	2.4 − 1	2.8 − 1	3.2 − 1	3.6 − 1
$M_{Ni}^s/M_0 \quad \xi = 2$	2.9 − 1	3.5 − 1	3.9 − 1	4.5 − 1
$n_e(255)$ cm^{-3}	1.16 + 7	5.41 + 6	3.24 + 6	1.91 + 6
$n_e(735)$ cm^{-3}	1.8 + 5	8.3 + 4	5.0 + 4	2.9 + 4
$\tau(78)$ gm cm^{-3} $\xi = 1$	6.57	5.12	4.30	3.60
$\tau(78)$ gm cm^{-3} $\xi = 2$	4.93	3.84	3.23	2.70
$\eta \quad \xi = 1$	5.21 − 1	4.36 − 1	3.82 − 1	3.32 − 1
$\eta \quad \xi = 2$	4.24 − 1	3.49 − 1	3.04 − 1	2.61 − 1
$\tau_{22.9}(435)$	4.50 + 1	2.10 + 1	1.26 + 1	7.42
$\tau_{22.9}(735)$	1.58 + 1	7.35	4.42	2.60
$\mathcal{E}_{IR}(435)$ erg s^{-1}	7.07 + 36	2.25 + 37	5.73 + 37	1.54 + 38
$\mathcal{E}_{IR}(735)$ erg s^{-1}	1.77 + 37	5.15 + 37	1.08 + 38	3.72 + 38

In Table II the volume of the shell and the energy deposited per cm^3 per sec at $t_d = 255$ d, $\mathcal{E}_d(255)$, are tabulated as a function of the expansion velocity, u.

The production rate, p^{+i}, of the i th stage of ionization can be determined from the energy deposition through the following considerations. The energy loss per cm path dx of electrons of energy E in a plasma per ion of type i is determined by the loss function,

$$L(E) = (1/N)(dE/dx) \tag{2}$$

as was discussed by Meyerott[17]. To a first approximation, the loss function should be independent of the stage of ionization. (It is the product of a cross section which varies inversely as the ionization potential and an energy loss per collision which is proportional to the ionization potential.) Hence, the energy loss to the i th ionic species per cm path is

$$(dE^i/dx) = (N^{+i}/N)(dE/dx) \tag{3}$$

The total energy lost by the slowing down of $\mathscr{F}$ electrons per cm^2 per sec of energy E is equal to the energy deposit per cm^3 per sec, $\mathcal{E}_d$. Since Equation 3 holds at all energies, the energy deposited in the i th ionic species, $\mathcal{E}_d^i$ is

$$\mathcal{E}_d^i = (N^{+i}/N)\mathcal{E}_d. \tag{4}$$

The production rate, p^{+i}, is given by the sum of the production rate due to energy loss of the primary electrons and the energy losses of the secondary, tertiary, etc., electrons produced in the slowing down process. Since the secondary electrons typically have a mean energy approximately twice the ionization potential of the species ionized, these electrons also can potentially cause additional ionization. As was discussed by Meyerott[17], the importance of the secondary electrons in producing ionization, in addition to that produced by the primary electrons, depends on the fractional ionization. In Figure 2 is shown the loss function for Cu, as a function of electron energy, computed from the stopping power of Cu, as given by Bichsel[18]. Also shown in Figure 2 is the loss function for free electrons as a function of electron energy for fractional ionization, $F = n_e/N$, of 1, 2, and 3, where n_e is the electron density. It can be seen in Figure 2 that the loss function of the free electron gas is equal to loss function for the ions at E ~ 500 eV. At an electron energy of 10^4 eV the loss function of the electron gas is lower than that for the ions by a factor of ~ 4, while at 10^6 eV it is down by a factor of ~ 25. Hence, a 10^6 eV electron in slowing down to 10^4 eV, and losing 99% of its energy, deposits practically all of its energy in the ions, while secondary electrons with energies up to ~ 500 eV will deposit practically all of their energy into the electron gas and be thermalized.

The fact that the primary electrons deposit all of their energy in the atomic ions, while the secondary electrons deposit their energy in the electron gas, greatly simplifies the problem of calculating the ion production rates and the ionic concentrations. The energy deposited in the electron gas is available for thermal ionization.

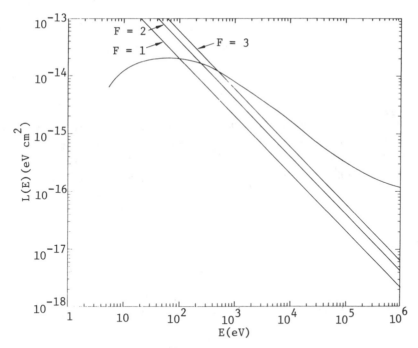

Fig. 2 Loss function for the free electrons gas for fractional
ionization 1, 2, and 3, and for Cu, as a function of
electron energy.[7]

However, since the synthetic spectral comparison with the observations indicate a low electron temperature, $T_e \sim 4000°$ K, little
ionization will arise from this source.

The ion production rates are now easily determined. As remarked
above, the mean energy of the secondary electrons is ~ 2 times the
ionization potential of the ionizing ions. If we add this to the
ionization potential of that species, the energy to produce an ion
pair, $w^i \cong 3I^{i-1}$. In this estimate is $\sim 10\%$ of the energy lost in
excitation. The ion production rate of the i th stage of ionization
is then

$$p^{+i} = \frac{N^{+(i-1)}\mathcal{E}_d}{3I^{i-1}N}$$ (5)

The steady state rate equations that determine the species
concentrations are for i > 0,

$$p^{+i} + \alpha^{+(i+1)}N^{+(i+1)}n_e - p^{+(i+2)} - \alpha^{+i+1}_N n_e = 0$$ (6)

$$N = \sum_{i=0} N^{+i} , \tag{7}$$

and

$$n_e = \sum_{i=1} (i)N^{+i} , \tag{8}$$

where N^{+i} is the concentration of the i th stage of ionization, N is the total species concentration, n_e is the electron density, and α^{+i} is the recombination coefficient for the i th stage of ionization. In order to have $dN^0/dt = 0$, the following equation must hold.

$$\alpha^+ N^+ n_e - p^1 = 0 \tag{9}$$

Substituting of (9) in (6) gives

$$N^{+i} = p^{+i}/(\alpha^{+i} n_e). \tag{10}$$

Substituting expression (5) for p^i, N^{+i} becomes

$$N^{+i} = \frac{N^{+(i-1)}(\mathcal{E}_d/N)}{3I^{i-1}\alpha^{+i}n_e} \tag{11}$$

The quantities n_e/N and N^i/N are a function of $N\mathcal{E}_d^{-\frac{1}{2}}$. This can be seen as follows: From relation (8) for n_e,

$$n_e/N = [\mathcal{E}_d^{\frac{1}{2}}/N]\left[\sum_{i=1} i(N^{+(i-1)}/N)/3I^{i-1}\alpha^{+i}\right]^{\frac{1}{2}} \tag{12}$$

and from (11) and (12)

$$N^{+i}/N = (\mathcal{E}_d^{\frac{1}{2}}/N)(N^{+(i-1)}/N)(1/3I^{i-1}\alpha^{+i})\left[\sum_{i=1} i(N^{+(i-1)}/N)/3I^{i-1}\alpha^{+i}\right]^{\frac{1}{2}} \tag{13}$$

It is clear that a set of values, n_e/N and N^{+i}/N, for given values of $\mathcal{E}_d$ and N will also be a solution for any other pair of values of $\mathcal{E}_d$ and N, provided $N\mathcal{E}_d^{-\frac{1}{2}}$ is constant.

Equations (7), (8), and (11) have been solved for n_e/N and N^{+i}/N, using the recombination coefficients for Mg ions at $T_e = 4000°$ K taken from Summers[19], $\alpha^+ = 5 \times 10^{-13}$, $\alpha^{+2} = 3 \times 10^{-12}$, $\alpha^{+3} =$

1×10^{-11}, and $\alpha^{+4} = 3 \times 10^{-11}$, all in units of $cm^3 s^{-1}$. In Figure 3 n_e/N and N^{+1}/N are shown plotted as a function of $N(\mathcal{E}_d/7.44 \times 10^3)^{-\frac{1}{2}}$. It can be seen that $N(\mathcal{E}_d/7.44 \times 10^3)^{-\frac{1}{2}} = 2.35 \times 10^6 cm^{-3}$ when $N^{++}/N^+ = 4$, the value for $t_d = 255$ d.

The scaling quantity, $N(t_d)\mathcal{E}_d(t_d)^{-\frac{1}{2}}$, can be shown to be related to the values at 255 d through the following relation,

$$N(t_d)\left|\mathcal{E}_d(t_d)/7.44 \times 10^3\right|^{-\frac{1}{2}} = 2.35 \times 10^6 \left[\frac{\mathcal{L}(255)}{\mathcal{L}(t_d)}\right]^{\frac{1}{2}} \left[\frac{255}{t_d}\right]^{3/2}. \quad (14)$$

$\mathcal{L}(t_d)$ is tabulated in Table I and is independent of the expansion velocity, u. The values of $N(t_d)\left|\mathcal{E}_d(t_d)/7.44 \times 10^3\right|^{-\frac{1}{2}}$ for $t_d = 78$ d, 435 d, and 735 d are also shown in Figure 3. It can be seen that the ratio of the twice ionized ions to the singly ionized ions is reduced to ~ 2.3 at 435 d and 0.5 at 735 d. In Figure 4a is shown the ratio of the energy radiated per sec in each 40 Å band to that radiated from 3320 Å to 11,600 Å, as a function of wavelength for $t_d = 435$ d. In Figure 4b is shown the synthetic spectrum,$0.6[\text{Fe II} + 2\text{Fe III}]$, as a function of wavelength calculated in the same manner as that for Figure 1b. It can be seen that there is fair agreement between the observed and synthetic spectra, considering the fact that at 435 d the electron density is ~ $10^6 cm^{-3}$, so that deviations from a Boltzmann population of levels might be expected to be important. However, the feature at 4600 Å does appear to be reduced relative to that at 4300 Å by about a factor of 2, compared to the spectra at 255 d shown in Figure 1a.

The latest observations were taken at $t_d = 735$ d when the SN was very weak and, hence, the records are quite noisy. However, it appears from the data shown in Kirshner and Oke[5] that the 4600 Å feature is greatly reduced or missing. This is consistent with the predicted ratio of N^{+2}/N^+ of ~ 0.5 shown in Figure 3.

At 78 d the predicted ratio, N^{+2}/N^+, is 2.9, which is lower than the factor of 4 predicted at 255 days. While the fraction of the energy in the feature at 4600 Å is down from that at 255 days, as can be seen in the data of Kirshner and Oke[5], the total Fe concentration may be lower at that time. If most of the Fe is freshly synthesized by the explosion, 1/2 of the final value should be present at 78 d, the half life of ^{56}Co. Because the spectra of ionized Co are essentially unknown, it is not possible to obtain an effective temperature, as was done at the later times for Fe. Hence, a comparison of the

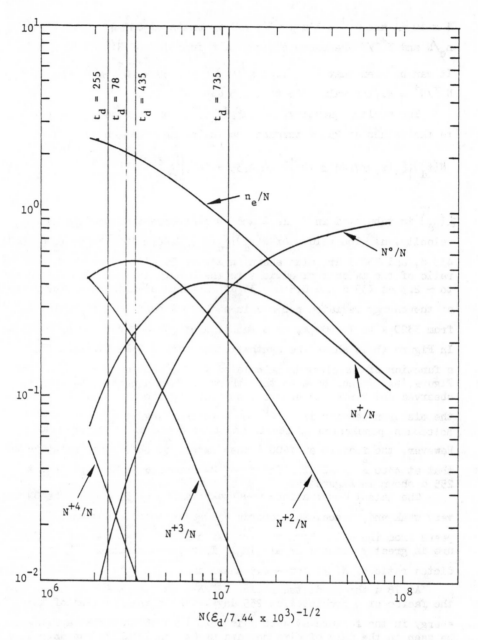

Fig. 3 The ratios n_e/N and N^{+i}/N as a function of $N(\varepsilon_d/7.44 \times 10^3)^{1/2}$

78 d spectra cannot be made at this time.

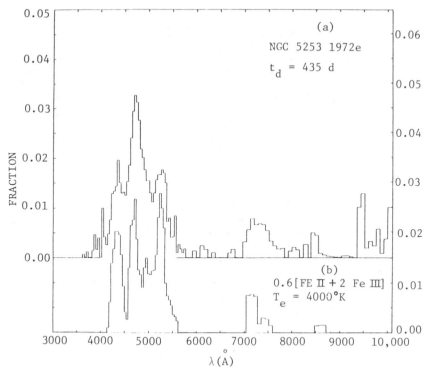

Fig. 4a Spectrum at t_d = 435 d versus wavelength plotted in same manner as Fig. 1a.

Fig. 4b Synthetic spectrum 0.6 Fe II + 2Fe III versus wavelength plotted in same manner as Fig. 1b.[7]

III. MASS OF THE SN ENVELOPE, MASS OF Fe, AND MASS OF Ni EXPLOSIVELY SYNTHESIZED

The ionic concentrations shown in Figure 3 can be used to calculate the mass of the SN envelope, the mass of Fe, and the mass of Ni explosively synthesized as a function of mean expansion velocity, u. Since we do not have a good measure of the expansion velocity, we shall tabulate these quantities as a function of u.

a) Mass of the SN Envelope

From the value of $\mathcal{E}_d(255)$ shown in Table II, the value of N can be calculated as a function of u. These values are shown in Table II. Also shown in Table II is the total number of atoms in the SN envelope, $\mathcal{N}$ = NV, where V is the volume. If we take N= ξN_{Fe}, where N_{Fe} is the number of Fe atoms/cm^3, and assume that the other atoms all have an atomic weight of 28, and mass of the envelope can be calculated as a function of u for a given value of ξ. In Table II are shown the masses of the envelope as a function of u for ξ = 1 and 2.

b) Mass of Fe in the SN Envelope

The mass of Fe in the SN envelope can be determined by calculating the amount of Fe required to produce the observed radiation rate at 255 d. In Figure 5 are shown the emissivities per ion of Fe II and Fe III plotted as a function of electron temperature, using a Boltzman distribution of levels. Also shown are the emissivities for Fe III at T_e = 5000° K as a function of electron density, n_e, calculated by Garstang, Robb, and Rountree [20]. In Figure 6 the emissivity of Fe III at T_e = 5000° K is shown plotted as a function of electron density. The emissivity of Fe IV is expected to be small, since the lowest excited level is at ~ 4 eV [21]. As indicated before, the fraction of the energy required to produce an ion pair that goes into the production of secondary electrons and, hence, into the electron gas, is 0.6. This is the fraction of the energy available for the excitation of forbidden line radiation in Fe II and Fe III. The remainder of the observed radiation should be the result of radiative recombination.

The number of iron atoms at a temperature of 4000° K required to radiate $0.6 \times 2.51 \times 10^{52}$ eV s^{-1}, the rate of energy radiated per sec at 255 d due to forbidden line cooling, can be estimated as follows. From Figure 5 it can be seen that the emissivity of Fe II and Fe III at 4000° K calculated using a Boltzmann population of levels is 3.4×10^{-3} and 7.8×10^{-4} eV s^{-1} ion^{-1}, respectively. If n_{Fe} is the total number of Fe atoms in the SN envelope, the number of Fe$^+$ and Fe^{++} ions at 255 d can be calculated, using the fractional abundance numbers from Figure 3, to be $1.4 \times 10^{-1} n_{Fe}$ and $5.2 \times 10^{-1} n_{Fe}$, respectively. The total energy radiated, assuming a Boltzmann population of levels, is $n_{Fe}[1.4 \times 10^{-1} \times 3.4 \times 10^{-3} + 5.2 \times 10^{-1} \times 10^{-4}]$ = $8.82 \times 10^{-4} n_{Fe}$ eV s^{-1}. If the reduction in the emissivity due to a non Boltzman population of levels is ζ, then the total energy radiated per second becomes $8.82 \times 10^{-4} \zeta n_{Fe}$ eV s^{-1}. The reduction factor, ζ, is a function of the electron density. Taking the electron densities at 255 d from Table II, ζ can be estimated using Figure 6, if we assume the reduction in emissivity for Fe II and Fe III at 4000° K are the same as that for Fe III at 5000° K shown in Figure 6. Equating the above expression for the predicted energy radiated per second by the forbidden lines of Fe$^+$ and Fe^{++} to $0.6 \times 2.5 \times 10^{52}$ eV s^{-1} allows us to evaluate n_{Fe}. The number of Fe atoms required is shown in Table II.

It can be seen from Table II that the total number of atoms of

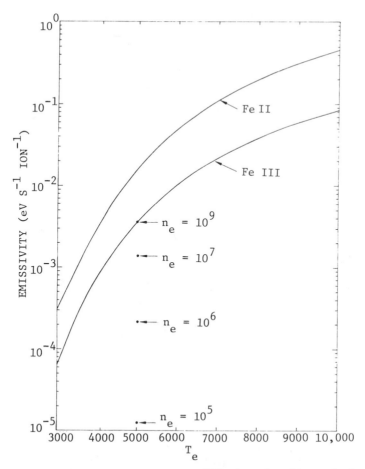

Figure 5 Emissivity of Fe II and Fe III as a function of electron
temperature for a Boltzmann distribution of levels, and
emissivity of Fe III at T_e = 5000° K as a function of
electron density. [7]

Fe required to produce 0.6 of the observed radiation rate at 255 d is about an order of magnitude larger than the total number of atoms in the envelope as predicted by the deposition code. In order for the total number of Fe atoms to be less than the total number of atoms in the envelope, the emissivity would have to be an order of magnitude larger, corresponding to an electron temperature of $\sim$ 5500° K, rather than $\sim$ 4000° K.

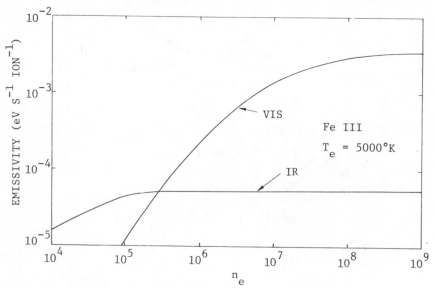

Figure 6   Emissivity of Fe III in the visible and the infrared at T_e = 5000° K as a function of electron density.[7]

The low value of the electron temperature determined from the spectral distributions in Figure 1 is probably indicative of interstellar absorption. Chiu et al[9] have suggested interstellar reddening corrections for NGC 5253. They indicate that the absorption correction at $\sim$ 4300 Å might be $\sim$ 50% larger than at 7200 Å. Since these are the two spectral regions which were used to determine the spectral temperature, an increase in the intensity of the 4300 Å band by 50% over the 7300 Å band would require an electron temperature close to 6000° K.

The absorption corrections and the change in electron temperature will have little effect on the ratio of Fe^{++} to Fe^{+} required to fit the data. This ratio was determined by making the intensities of the predicted spectra at 4300 Å and 4600 Å agree approximately with the observed intensities. Since these two features arise from levels in Fe^{++} and Fe^{+} having approximately the same excitation potentials, the Fe^{++} to Fe^{+} ratio should be insensitive to small changes in electron temperature.

The absorption correction of the magnitude suggested by Chiu et al[9] has the effect of increasing the luminosities in Table I and by a factor of ~ 2. Since the atom density N for a constant N^{+i}/N_d ratio varies as $\epsilon_d^{\frac{1}{2}}$, the factor of 2 change in luminosity would change η (255) and the mass of the envelope indicated in Table II by $\sim 40\%$. An uncertainty of this magnitude probably exists in the observed luminosities due to the uncertainty in the distance scale.

c) Mass of Ni Explosively Synthesized

The mass of Ni explosively synthesized can be calculated by requiring the rate of energy deposition in the SN envelope by positron and γ-rays to equal the total observed rate of radiation. Since the half life of ^{56}Ni is 6 d, for times of 78 d or longer, practically all of the ^{56}Ni will have decayed into ^{56}Co. ^{56}Co decays partly by electron capture with the emission of γ-rays and partly by positron decay. The mean decay energy is ~ 3.6 MeV per decay with ~ 0.96 of the energy as ~ 1 MeV γ-rays and ~ 0.04 of the energy emitted as positrons with a mean energy of ~ 0.64 MeV. The rate of deposition of energy depends on the mass density of the shell as well as the energy decay rate.

The thickness of the shell, in grams/cm^2, is equal to the mass density times the shell thickness, 0.2R. In Table II is shown the thickness of the shell at 78 d, τ(78), as a function of u. The range of a 0.64 MeV positron is only ~ 0.3 gm/cm^2. Since $\tau(78) \simeq 5$ gm/cm^2, all the positron energy is deposited in the envelope. MeV γ-rays lose energy by Compton scattering. The mass absorption coefficient for Compton scattering, μ/ρ, for MeV γ-rays is 2.54×10^{-1} cm^2/gm. In scattering the γs lose energy to the electrons. The mass absorption coefficient for energy loss is 1.12×10^{-1} cm^2/gm. Since the thickness of the shell at 78 d is $\sim.5$ gm/cm^2, only single scattering need be taken into account. In Table II the fraction of the γ-ray energy deposited in the SN, η, is shown tabulated as a function of expansion velocity, u. The luminosity of the SN at $t_d = 78$ d can be related to the total number of Ni atoms explosively synthesized, η^s_{Ni}, as

$$\mathcal{L}(78) = [9.6 \times 10^{-1}\eta + 4.0 \times 10^{-2}]6.5 \times 10^{-13}\eta^s_{Ni}\exp(-78/111). \quad (15)$$

Equating $\mathcal{L}(78)$ to the observed luminosity at 78 d, 8.99×10^{41} erg s^{-1}, from Table I, we obtain the total number of Ni atoms synthesized

as a function of u. The mass of Ni synthesized, M_{Ni}^s, as a function
of u is shown in Table II. Since the value of η depends slightly on
the fraction of the atoms that are Fe, the mass of Ni synthesized
depends slightly on ξ, as shown in Table II. It can be seen in
Table II that the mass of the SN envelope calculated, using the atom
concentration derived from the deposition code, is equal to or greater than the mass of Ni synthesized for values of u ≥ 5. If the absorption corrections discussed in § III, b), are applied, the mass of
Ni synthesized would be about equal to or greater than the mass of
the SN envelope calculated from the deposition code for u ≥ 10.

IV. INFRARED EMISSION

Some energy should be radiated in the infrared spectral region
at late times if most of the observed visible radiation is due to
Fe II and Fe III as suggested in the previous sections. The fraction
of the total emission rate radiated in the IR should increase with
time. Since we have taken the visible emission rate to be a measure
of the rate of energy deposition by positrons and γ-rays, the energy
deposition used in this paper could be too small if there is a large
fraction of the emission in the IR.

IR emission should occur from transitions between the levels in
the ground terms of Fe I, Fe II, and Fe III. The ground term of
Fe IV is a singlet, so that there will be no contribution to the IR
from Fe IV. The excitation potentials of these levels are ~ 0.1 eV,
so that they are readily excited at temperatures of 4000° K. Hence,
in spite of the fact that the transition probabilities are ~
10^{-3} s^{-1}, radiation from these levels can be important at small electron densities. In Table III are shown the wavelengths and transition probabilities of the strongest transitions in Fe I, Fe II, and
Fe III. The transition probabilities for Fe I are from Grevesse,
Nussbaumer, and Swings[22], those for Fe II and Fe III are from
Garstang[23]. It can be seen in Table III that the wavelengths of the
lines and the corresponding transition probabilities are almost the
same for Fe I, Fe II, and Fe III. Hence, the IR emission rate will
be essentially independent of the state of ionization of Fe and depend only on the total number of Fe atoms present in the SN envelope.
In this paper we shall only estimate the IR emission rate due to
Fe III.

In Figure 6 the sum of the emissivities of the 4 IR lines of
Fe III are shown plotted as a function of electron density for T =
5000° K. The data are from Garstang et al[20]. It can be seen that
the emissivity in the IR is essentially constant for electron densities $> 10^5$ cm^{-3}. It can also be seen in Figure 6 that the emissivity in the IR exceeds that in the visible for Fe III for electron
densities $< 3 \times 10^5$ cm^{-3}. The electron densities at $t_d = 435$ d and

735 d are $\sim 5 \times 10^5$ cm^{-3} and 5×10^4 cm^{-3}, respectively. Hence, we can expect the IR to make some contribution to the total emission rate at late times. The magnitude of the contribution will depend on the optical depth for the IR radiation.

In Table II are shown the optical depths, $\tau_{22.9}(435)$ and $\tau_{22.9}$ (735) for the 22.9 μm line as a function of expansion velocity, u, for t_d = 435 d and 735 d. It can be seen that the optical depths are greater than 1 for all values of u listed. The optical depths at earlier times are even larger than those shown due to the higher atom density. The IR emissivities shown in Figure 6 are constant for electron densities $> 10^5$ cm^{-3} because the collision rate is sufficiently high for the levels to have a Boltzman population. This also implies that the electron quenching rate exceeds the radiation rate. Under these conditions the probability of a photon being lost while being scattered is ~ 1. Hence, the optical depths shown in Table II can be assumed to correspond to photon loss. Thus, the only radiation that escapes is that which is emitted at optical depths less than 1. An estimate of the IR emission rate at 735 d, $\varepsilon_{IR}(735)$, is

$$\varepsilon_{IR}(735) = [1/\tau_{22.9}(735)][E^{IR}(735)]\eta_{Fe} \qquad (16)$$

where $E^{IR}(735)$ is the IR emissivity of Fe III shown in Figure 6. We use the optical depth for $\lambda = 22.9$ μm, since this line makes the largest contribution to the emissivity. The exact prediction of the IR emissivity needs more careful consideration. In Table II are shown the electron densities at 735 d as a function of u from which the value of $E^{IR}(735)$ can be obtained from Figure 7. The values of the emission rate at 735 d are shown in Table II as a function of u. Since we do not have a good estimate of η_{Fe}, η_{Fe} was taken to equal η for the purpose of this estimate.

The value of the luminosity in the visible at 735 d, shown in Table I, is 7.87×10^{37} erg s^{-1}. It should be kept in mind that this value is only extrapolated from the others listed in Table I and may be in error by more than that due to the error in the distance scale. It can be seen that the estimated value for the IR emission rate is about the same as the extrapolated emission rate in the visible. An upper limit to the IR emission rate would be that given by black body emission over the widths of the lines. This limit would be about a factor of 10 larger than that given in Table II at 735 d but about the same as that at 435 d. Both estimates might also be increased by a factor of 2 if the emission from the inside of the receding portion of the expanding shell can escape through the approaching portion. If the IR emission is as large at 735 d as these estimates imply, the energy deposition rate at this time should be increased. In which case, a larger contribution to the 735 d spectra from Fe III would be predicted.

The emission rate in the IR for 435 d, $\varepsilon_{IR}(435)$, is also shown in Table II, calculated in the same manner as that at 735 d. Since the electron densities at 435 d are all greater than 10^5 cm^{-3}, the emissivity does not depend on n_e. the luminosity in the visible at 435 d from Table I is 4.20×10^{39} erg s^{-1}. It can be seen that this is more than an order of magnitude larger than the emission rate in the IR at 435 d shown in Table II. The IR will contribute even less to the total luminosity at earlier times. Hence, taking the rate of deposition of energy as equal to the luminosity in the visible shouls be a good approximation at 78 d, 255 d, and 435 d.

TABLE III

WAVELENGTHS AND TRANSITION PROBABILITIES
OF THE GROUND TERM TRANSITIONS
IN Fe I, Fe II, AND Fe III

Transition	$\lambda(\mu m)$	$A(s^{-1})$
Fe I a^5D–a^5D		
4–3	24.0	2.5×10^{-3}
3–2	34.7	1.6×10^{-3}
2–1	54.3	6×10^{-4}
Fe II a^6D–a^6D		
4–3	26.0	2.1×10^{-3}
3–2	35.4	1.6×10^{-3}
2–1	51–3	7.2×10^{-4}
1–0	87.4	1.9×10^{-4}
Fe III a^5D–a^5D		
4–3	22.9	2.8×10^{-3}
3–2	33.0	1.8×10^{-3}
2–1	51.7	6.7×10^{-4}
1–0	105.	1.4×10^{-4}

III. MECHANISMS FOR THE EXCITATION
OF LATE TIME SPECTRA OF TYPE II SUPERNOVAE

Type II supernovae (SN II) are thought to be considerably more massive than SN I with a composition that is more nearly solar. The evolution and explosion model of Weaver and Woosley[3] and Weaver, Zimmerman, and Woosley[24] with a pre-explosive mass of 15 M$\odot$ and 0.1 to 0.4 M$\odot$ of explosively generated ^{56}Ni is shown to be a fair fit for the light curve of SN II 1969ℓ. Most of the energy to power the light curve after 100 day in their model is due to the radio-active decay of ^{56}Co. Their calculations indicate the presence of four distinct zones: 1) a zone of $\sim$ 0.2 M$\odot$ of predominately Ni, Si, S, Mg, and O; 2) a zone of $\sim$ 0.8 M$\odot$ of predominately Ne and O; 3) a zone of $\sim$ 2 M$\odot$ of He; and 4) a zone of $\sim$ 11.5 M$\odot$ of predominately H and He.

The radioactive energy is generated in the first zone containing the ^{56}Ni and its decay products. Until this zone becomes transparent to γ-rays, all of the radioactive energy will be deposited in this zone. When the γ-rays start to escape from the heavy metal zone, some of the energy will be deposited in the more massive outer zones. In spite of their greater mass, these outer zones may also be partly transparent to γ-rays at the time that the Ni zone becomes transparent because of higher relative expansion velocities and, hence, lower densities. The fact that the Balmer lines of H appear to increase in importance relative to the rest of the spectra for 1969 as a function of time may be indicative of this increased energy deposit with time. While part of this increase in prominence of the Balmer lines may be due to increased optical transparency, it may also be due partly to the increase in the γ - deposition in the H-He zone at the later times. The absence of any prominent H or He lines in the late time spectra of SN I is probably due to the absence of any massive zone of H or He outside the Ni zone.

A rough estimate of the late time ionization state of 1969ℓ can be made by assuming a 15 M$\odot$ envelope with solar composition with a radioactive deposition rate equal to the observed luminosity at late times. At 200 days the luminosity is $\sim$ 2 x 10^{41} erg/sec.[3] Assuming an expansion velocity of 10^9 cm/sec, the atom density N, the electron density n_e, and the fractional ionization, F, are $\sim$ 8.7 x 10^8 cm^{-3}, $\sim$ 1.6 x 10^7 cm^{-3}, and 1.8 x 10^{-2}, respectively. In the estimate for the electron density, the energy to produce an ion pair in H was taken to be 60 eV and the recombination coefficient = 4 x 10^{-13} cm^3 s^{-1}. In contrast to the situation in SN I, it would appear that the late time SN II envelope is only weakly ionized.

The details of the energy deposition in the Ni zone can be expected to differ somewhat from that of the SN I Ni zone. In Figure 7 is shown the loss function for the energy loss of electrons to the

electron gas for fractional ionization, $F = 1.8 \times 10^{-2}$, and the loss
function of Cu. It can be seen that only electrons with $E < 10$ eV
deposit their energy into the electron gas. This means that a smaller fraction of the energy from the secondary electrons will be deposited in the electron gas than the corresponding case for SN I.
With the lower average fractional ionization the forbidden lines of
Fe I may make a contribution to the radiative cooling. The Mg I
lines may also be formed in this region where the relative Mg concentration is large.

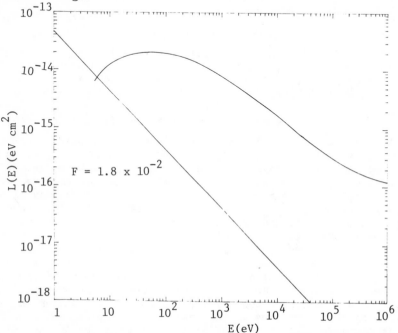

Figure 7   Loss function for the free electron gas for fractional
ionization 1.8×10^{-2}, and for Cu, as a function of
electron energy.[7]

In the massive H-He outer zone most of the radioactive energy
will be deposited in H and He. The loss functions $L_H(E)$ and $L_{He}(E)$
for hydrogen and helium respectively, have been determined by Green
and Peterson[25]. The contribution of H and He to the total loss function as a function of electron energy, E, for a mixture with solar
composition of 94% H and 6% He is shown in Figure 8. Also shown in
Figure 8 is the loss function of the free electrons for a fractional
ionization, $F = 1.8 \times 10^{-2}$. The energy loss to the metals is much
smaller than the loss to the free electrons at all energies. It can
be seen in Figure 8 that MeV electrons, in slowing down to $\sim 10^4$eV
will deposit $\sim 90\%$ of their energy in H and $\sim 10\%$ in He. Secondary

electrons with energies from $\sim$ 40 eV to 500 eV will lose most of that energy to H or the electron gas. Secondary electrons with energies below $\sim$ 40 eV will deposit their energy into the electron gas.

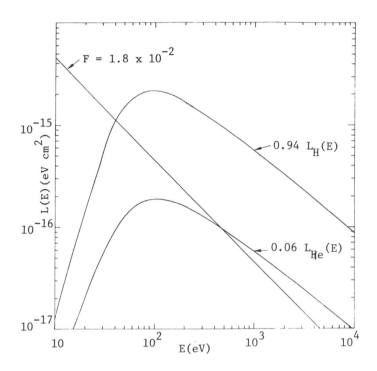

Figure 8 Contribution of H and He to the total loss function for a number composition of 94% H and 6% He, and the loss function for the free electron gas for fractional ionization of 1.8 x 10^{-2}, as a function of electron energy.

In Table IV is shown the fraction of the primary energy that results in ionization, excitation, secondary electrons, the mean energy of the secondary electrons, and the energy required to produce an ion pair in H and He. The H data is from Bethe[26] and the He data is from Peterson and Green.[27] In Table V is shown the fraction of the primary energy deposited in the excited states of H. The mean energy of the secondary electrons is 31 eV and 50 eV, respectively. Hence, it can be seen in Figure 8 that about one half of the energy of the secondary electrons from H is deposited in the electron gas. The energy of the secondary electrons from He will be deposited either in H or the electron gas. Since we have assumed an abundance of He of 6%, the secondary electron contribution from He will be small. The following conclusion concerning the excitation and ionization of H and He can be drawn at this point:

1) The excitation and ionization of H will be principally by primary electrons.

2) Both electron collisional excitation and radiative recombination will contribute to the population of the excited states of H.

3) The excitation and ionization of He will be by the primary electrons. Only the singlet states will be excited.

4) The initial electron energy deposition produces little or no excitation of the metals.

5) Approximately 50% of the energy deposit is in the electron gas which is the source of the thermal excitation.

6) Approximately 30% of the energy deposit will be radiated by H, principally in $L\alpha$.

7) Approximately 20% of the energy deposited is in H ionization, H^+.

TABLE IV

ENERGY PARTITION OF PRIMARY ELECTRONS

	H	He
Ionization	0.21	0.29
Excitation	0.32	0.12
Secondary Electron	0.48	0.59
E_s(eV)	31	50
W_p(eV)	67	85

TABLE V

FRACTION OF PRIMARY ENERGY DEPOSITED
IN EXCITED STATES OF H

n	Energy Fraction
2	0.237
3	0.047
4	0.017
5	0.008
> 5	0.015
All Discrete	0.32

Since the metals cannot be ionized by the primary electrons and the temperature of the electron gas appears to be too low to produce much collisional ionization, some other sources of ionization must exist if there is to be appreciable metal ionization. Two possible sources of ionization of the metals are the ultraviolet radiation from H and He and charge transfer ionization between the H^+ and He^+ ions and the metal atoms. In this report we shall consider only the charge transfer possibilities. The following charge transfer reactions are likely to be important ionization and excitation mechanisms in a plasma having predominately H^+ ions:

$$H^+ + O(2p^4\ {}^3P) \longrightarrow H(1s\ {}^2S) + O^+(2p^3\ {}^4S^\circ) \tag{17}$$

$$H^+ + Si(3p^2\ {}^3P) \longrightarrow H(1s\ {}^2S) + Si^+(3p^2\ {}^4P) + 0.11 \tag{18}$$

$$H^+ + S(3p^4\ {}^3P) \longrightarrow H(1s\ {}^2S) + S^+(3p^3\ {}^2P^\circ) + 0.21 \tag{19}$$

$$H^+ + Ca(4s^2\ {}^1S) \longrightarrow H(1s\ {}^2S) + Ca^+(5\ {}^2P^\circ, 4\ {}^2D) + (0.01, 0.49) \tag{20}$$

$$H^+ + Fe(a\ {}^5D) \longrightarrow H(1s\ {}^2S)$$
$$+ Fe^+(z\ {}^4P^\circ, z\ {}^4D^\circ, z\ {}^4F^\circ, z\ {}^6P^\circ, z\ {}^6F^\circ, z\ {}^6D^\circ) + \Delta E \tag{21}$$

Of special interest are the reactions (19), (20), and (21) which produce S^+, Ca^+, and Fe^+. These ions all have low-lying metastable levels which can be excited by electron collisions at temperatures as low as 5000° K.

TABLE VI

SPECTRA AND EXCITATION MECHANISMS IN THE H - He ZONE

Spectra	Excitation Mechanism
H Balmer	Primary Electrons, Secondary Electrons Recombination
He Singlet Triplet	Primary Electrons, Recombination Recombination
S II (1F)	Charge Transfer, Thermal
Ca II (3), (4), (5) (1), (2), (1F)	Charge Transfer Charge Transfer, Thermal
Fe II (37), (34), (42) Forbidden Lines	Charge Transfer (Charge Transfer), Thermal
CI (3F), OI (3F) Mg I(1), Si I (1F) SI (1F), (2F), (3F) Fe I Forbidden Lines	Thermal

In Table VI are shown a list of the expected spectra and the excitation mechanisms in the H-He zone. H is excited by primary and secondary electrons. He singlet lines are excited by primary electrons and recombination. The He triplet lines, if any, must be excited by recombination. The S^+, Ca^+, and Fe^+ ions can be excited by charge transfer. The low-lying levels are excited by electron collisions from the electron gas. Permitted line radiation terminating on excited states may escape. Radiation from metastable levels with life times shorter than 10 - 100 s will probably escape. Forbidden line radiation from the low-lying levels of C, O, Si, S, Fe, and the 4571 Å line of Mg will be excited thermally by the electron gas. Most of this radiation should escape.

IV. SUMMARY

In summary, it has been shown that energy deposition in the envelopes of SN at late times by positrons and γ-rays is initially deposited in the most abundant species and in the electron gas. The high energy primary and secondary electrons deposit their energy into the atoms and ions and determine the initial species concentrations of the most abundant species. In SN II charge transfer collisions may determine ionization of the less abundant elements. The lower energy secondary electrons deposit their energy in the electron gas. The energy deposited in the electron gas may amount to 50% or more of the total energy deposit. The temperature of the electron gas is determined by the rate of energy deposit and the rate of radiation cooling. Forbidden line radiation by Fe^+ and Fe^{++} appears to be the principal cooling mechanism for late time SN I. While no quantitative analysis has been made for SN II, it seems likely that forbidden line radiation by C, O, Si, S, Fe, Mg, S^+, Ca^+, and Fe^+ will also be the principal cooling mechanism for late time SN II.

The production of ionization by MeV electrons in a low density gas is more efficient ionization process than thermal ionization. In the thermal ionization process a larger amount of energy goes into excitation for each ion pair produced than is the case for ionization by MeV electrons. Consequently, recombination radiation makes only a small contribution to the total radiative energy loss rates[28]. In the late time SN envelopes it would appear that as much as 30 to 50% of the energy deposited is in ionization. Hence, the recombination radiation is expected to make an appreciable contribution to the late time spectra.

Further progress in the improvement of the radioactive source model for the late time spectra of SN is handicapped by the lack of some of the atomic physics data. The most important data required are the following:

1) Energy levels and transition probabilities for Ni to Ni^{+4}, Co to Co^{+4}, and Fe to Fe^{+4}.

2) Low temperature low electron density emissivities of Fe, Fe^+, Fe^{+2}, Co, Co^+, and Co^{+3}.

3) Recombination coefficients and spectra of Ni^+ to Ni^{+4}, Co^+ to Co^{+4}, and Fe^+ to Fe^{+4}.

4) Loss functions for the metals Si, Mg, S, Ca, Fe, Co, and their ions to at least the 4th stage of ionization; the determination of the energy to produce an ion pair in atomic ions, w^i; the fraction of the electron energy loss that goes into excitation and ionization as a function of fractional ionization.

5) Cross sections for charge and energy transfer between H^+ and, possibly, He^+, and the metals.

ACKNOWLEDGMENTS

The author would like to thank Dr. H. M. Johnson and the other members of the Astrophysics group of the Lockheed Palo Alto Research Laboratory for many helpful discussions. Thanks are especially due to Dr. A. Petschek for discussions of the IR emission. This research was supported (in part) by Independent Research and Development funds of the La Jolla Institute.

REFERENCES

1. S. A. Colgate and C. McKee, Astrophys. J. 157, 623 (1969).
2. W. D. Arnett, Astrophys. J. Lett. 230, L37 (1979).
3. T. A. Weaver and S. E. Woosley, Preprint UCRL-82057, Rev. (1978).
4. S. A. Colgate, A. G. Petschek, and J. L. Kriese, Preprint (1979) (See also this volume.)
5. R. P. Kirshner and J. B. Oke, Astrophys. J, 200, 574 (1975).
6. R. P. Kirshner and J. Kwan, Astrophys. J., 197, 415 (1975).
7. R. E. Meyerott, to be published Astrophys. J. (1980).
8. C. Gordon, Astrophys. J. 216, 67 (1977).
9. B. C. Chiu, P. Morrison, and L. Sartori, Astrophys. J. 198, 617 (1975).
10. M. Abdulwahab and P. Morrison, Astrophys. J. 220, 1087 (1978).
11. J. B. Oke, Private comm. (1978).
12. J. L. Sersic, G. Carranza, and M. Pastoriza, Astrophys. and Space Sci. 19, 469 (1972).
13. A. V. Holm, C. C. Wu, and J. J. Caldwell, Publ. Astron. Soc. Pac. 86, 296 (1974).
14. R. P. Kirshner, S. P. Willner, E. E. Becklin, G. Neugebauer, and J. B. Oke, Astrophys. J. Lett. 180, L97 (1973).
15. R. E. Meyerott, Bull. Am. Astron. Soc. 10, 638 (1978).
16. G. Lasher, Private Comm. (1979). (See also this volume.)
17. R. E. Meyerott, Astrophys. J. 221, 975 (1978).
18. H. Bichsel, Amer. Inst. of Phys. Handbook, 3rd Ed. (McGraw-Hill, N. Y.), P. 8-142 (1972).
19. H. P. Summers, Mon. N. R. Astron. Soc. 169, 663 (1974).
20. R. H. Garstang, W. D. Robb, and S. P. Rountree, Astrophys. J. 222, 384 (1978).

21. R. H. Garstang, Mon. N. R. Astron. Soc. <u>118</u>, 572 (1958).
22. N. Grevesse, H. Nussbaumer, and J. P. Swings, Mon. N. R. Astron. Soc. <u>151</u>, 239 (1971).
23. R. H. Garstang, Mon. N. R. Astron. Soc. <u>117</u>, 393 (1957); Mon. N. R. Astron. Soc. <u>124</u>, 321 (1962).
24. T. A. Weaver, G. B. Zimmerman, and S. E. Woosley, Astrophys. J. <u>225</u>, 1021 (1978).
25. A. E. S. Green and L. R. Peterson, J. Geophys. Res. <u>73</u>, 233 (1968).
26. H. Bethe, Handb. Phys. 2nd Ed. xxiv/I, 519 (1933).
27. L. R. Peterson and E. E. S. Green, J. Phys. B <u>1</u>, 1131 (1968).
28. V. L. Jacobs, J. Davis, J. E. Rogerson, and M. Blaha, Astrophys. J. <u>230</u>, 627 (1979).

RECENT ADVANCES IN CHARGED PARTICLE ENERGY DEPOSITION AND APPLICATIONS TO SUPERNOVA SPECTRA

A. E. S. Green

University of Florida, Gainesville, FL   32611

ABSTRACT

Most studies of charged particle energy deposition deal with
the stopping of particles by the medium. Predicting the excitation
of the medium by the particles has received much less attention.
The key to the solution of this latter problem is the assembly of a
realistic set of detailed atomic cross sections (DACS) for all important inelastic processes induced by electrons with energies varying from the incident electron energy down to the lowest inelastic
threshold. Two methods of assembling such sets are described; one
based upon experimental generalized oscillator strengths, the second
based upon quantum mechanical calculations using an atomic independent particle model. Next we describe a continuous slowing down
approximation (CSDA) method of utilizing these cross sections for calculating the excitation of a medium by incident electrons. We also
describe the yield spectra method based upon a modified discrete
energy bin technique which allows for the discrete nature of the
slowing down process. In addition, we describe a method of spatial-
yield spectra calculated by Monte Carlo techniques which gives the
spatial patterns of various excitations. Finally, we discuss the
possibility of applying these recent techniques to the prediction of
late time SN I spectra, using the radioactive excitation source model.

I.   INTRODUCTION

The radioactive decay model of late time supernova spectra calls
into play our understanding of beta particle and gamma ray energy
deposition. Studies of energy deposition by atomic particles or rays
date back over a century ago to the very beginnings of modern atomic
physics with work by Plücker, Hertz, Crookes, Goldstein, Thomson et
al. Among the most influential theoretical works during the early
years of this century were the continuous slowing down approximation
(CSDA) quasi-classical model of Bohr[1] and the CSDA quantum mechanical model of Bethe[2]. These works and most earlier and subsequent
studies mainly addressed the problem of quantifying the stopping of
charged particles by the medium. The problem of quantifying the excitation of the medium by the charged particles has received much
less attention. Indeed, it is only in the last twenty-five years
that sufficient atomic cross sections have become available to use
as basic inputs for excitation calculations. Once a set of detailed
atomic cross sections (DACS) has been assembled a variety of energy
apportionment techniques may be used to calculate the number of
various states of the medium which are excited as the initial particle and all secondary, tertiary, etc. electrons degrade in energy.
Table I summarizes early and recent calculational techniques which
can be used with DACS.

Table I. Energy Apportionment Techniques Usable with DACS

Population Equation (1922) (Fowler)[3]

Degradation Spectra (1954) (Spencer and Fano)[4,5]

Monte Carlo (1963) (Berger)[6]

CSDA Successive Generation Method (1965) (Green and Barth[7], Khare[8])

CSDA Integral Equation (1969) (Peterson and Green)[9]

Discrete Energy Bin (1969) (Peterson)[10]

Equilibrium Flux (1971) (Jura[11], Dalgarno et al.[12,13])

Scaled Degradation Spectra (1975) (Douthat[14], Fano[15])

Yield Spectra (1977) (Green, Jackman, Garvey, Porter)[16-18]

Variational Degradation Spectra (1978) (Rau, Inokuti, Douthat)[19]

The works listed in Table 1 may be divided into the CSDA methods which essentially extend and adapt the Bohr and Bethe models and all the others which allow for the discrete energy loss nature of the electron slowing down process.

Comparisons of the predictions of several of these apportionment methods, based on the same set of cross sections to assess the relative reliability and utility of each method, have been carried out by Garvey and Green[20]. Their main test consisted of a comparison of the values of W, the number of eV per ion pair, obtained from each method using molecular hydrogen as the medium. They actually compared four of these energy-apportionment methods, the CSDA integral equation, the Fowler equation, the discrete energy bin method and the Monte Carlo method. However, in effect, all nine methods were compared. The successive-generation CSDA approach is automatically summed by the integral equation CSDA approach. It has also been shown[14] that the degradation spectrum method leads to the same results as the population equation. The equilibrium flux and yield spectrum may both be derived from the discrete energy bin method[16]. The equilibrium flux has been shown to be equivalent to the degradation spectra[18].

The values of W predicted by the various discrete-loss energy apportionment methods are found to be in good agreement with one another. The costs of the calculations, however, vary considerably with the Monte Carlo method, generally the most costly. The CSDA method, the most economical method of calculation, gave W results higher than those of the discrete method.

While all of the discrete energy loss methods intrinsically require large computer codes and lack the simplicity and ease of application of the CSDA method, one of them, the yield spectrum method[16-18], appears to have advantages in applications. Once obtained by computer calculation it is amenable to accurate analytic representation, and is approximately invariant from substance to substance. Thus, it can conveniently be applied at low calculational costs to the approximate determination of all types of spectral yields.

In this paper we shall explore the possibility of applying the CSDA and the yield spectrum method to the explanation of late time supernova spectra. We shall be particularly concerned with excitations induced by incident electrons. Calculation of the excitations induced by gamma rays follows similar methods after allowing for pair production, the Compton effect, and the photoelectric effect.

II. THE DETAILED ATOMIC CROSS SECTION (DACS) APPROACH

The key to the determination of the excitation of the medium caused by electron deposition in gases is the initial assembly of a realistic set of detailed atomic cross sections (DACS) for the various inelastic processes induced by electrons whose energies vary from the incident energy all the way down to the lowest inelastic threshold. Of particular importance are differential ionization cross sections $S(E_O,T)$ where E_O is the primary electron energy and T the secondary electron. These secondary electrons mostly have low energies (i.e., $T < 50$ eV), energies which lie below the range of applicability of the Born plane wave approximation. Hence the Bethe-Born approximation (BBA), the basic pillar of almost all quantum mechanical approaches to stopping theory, must be supplemented or be surplanted by realistic treatments of low energy electrons in excitation theory. Low energy electrons, for example, are quite efficient in exciting forbidden emissions.

The detailed atomic cross sections approach can also utilize the BBA but in a different way than it is used in the treatment of stopping. Thus, at the University of Florida (UF), the BBA is used as a tool for the assembly of quantitative analytic representations of allowed and forbidden excitation cross sections and of differential and total ionization cross sections. The initial strategy in this program was dictated by the scarcity, at that time, of direct experimental cross section data but the availability of a substantial body of discrete and continuum generalized oscillator strength (GOS) data through the landmark work of Lassettre and his collaborators[21] which can be used to infer cross sections.

Lassettre carried out an extensive series of energy loss measurements for 500 eV incident electrons impacting upon various atmospheric gases and scattering through angles ranging up to 12° from the incident beam. By casting these data into a BBA framework Lassettre assigned GOS's for a wide variety of excitations and ionizations. Green and Dutta[22] then developed a semi-empirical procedure for analytically extrapolating these GOS to cover a broad range of momentum transfers. They also allowed for reasonable "distortions" of these generalized oscillator strengths at lower energies which in effect carried them outside the domain of the BBA.

The Born approximation as applied to inelastic electron collisions is well described in the literature (Mott and Massey[23]).
We use the Bohr radius, $a_o = \hbar^2/me^2 = 0.5292 \times 10^{-8}$ cm, as a unit of
length and the Rydberg energy, $R_e = \hbar^2/2ma_o^2 = me^4/2\hbar^2 = 13.602$ eV,
as a unit of energy. A dimensionless momentum squared transfer
parameter is introduced, which is defined by

$$x = a_o^2 K^2 = \frac{2E}{R_e} [1 - \cos\theta(1- \frac{W}{E})^{\frac{1}{2}} - \frac{W}{2E}] \tag{1}$$

where $K = k - k'$ is the difference between incident and outgoing
propagation vectors and $W = E_n - E_o$ is the excitation energy of the
nth state with respect to ground state. The minimum x (denoted
by x_ℓ) occurs for scattering in the forward direction ($\theta = 0$). The
maximum (x_u) occurs for backward scattering ($\theta = \pi$). From Eq. (1)
it is seen that as $E \to W$, $x \to x_t = W/R_e$ for all angles.

Using time-dependent perturbation theory in conjunction with
the Born approximation it can be shown that the differential cross
section for a collision leading to the excitation of the nth state
of an atom containing Z orbital electrons is

$$\frac{d\sigma}{d\Omega} = \frac{4I^2}{a_o^2 K^4} \frac{k'}{k} \tag{2}$$

where I is the matrix element of exp i $K \cdot r_i$ with respect to the
initial and final state and summed over all the atomic electrons.
The so-called generalized oscillator strength GOS is defined by

$$f(x) = (x_t/x) I^2 \tag{3}$$

This GOS goes over to the usual optical oscillator strength in the
limit $K \to 0$. The differential cross section can be expressed in
terms of GOS using

$$\frac{d\sigma}{d\Omega} = \frac{4a_o^2}{x_t} \frac{k'}{k} \frac{f(x)}{x} = \frac{4a_o^2}{x_t} (1- \frac{W}{E})^{\frac{1}{2}} \frac{f(x)}{x} \tag{4}$$

The total cross section (i.e.) the excitation cross section integrated over all angles of scattering is given by

$$\sigma = \frac{q_o}{WE} \int_{x_\ell}^{x_u} \frac{f(x)}{x} dx \tag{5}$$

where $q_o = 4\pi a_o^2 R_e^2 = 6.514 \times 10^{-14}$ eV^2 cm^2.

Green and Dutta calculated a set of analytic GOS's using helium
wave functions, the simplest system for which experimental data was
available. These analytic f(x) are functionally quite complex and
difficult to integrate. For applications, Green and Dutta found it
convenient to approximate these functions by the integrable
analytic form

$$f(\xi) = \sum_{s=1}^{N} f_s \, \xi^s \, \exp{-\alpha_s \, \xi} \tag{6}$$

where $\xi = x/x_t$ and f_s and α_s are adjustable parameters. This form has been successful in fitting Lassettre's data, the helium theoretical results, as well as results obtained using the independent particle model described in the next section.

The integrated cross section corresponding to Eq. (6) is

$$\sigma = \frac{q_o}{WE} \quad f_o[E_1(\alpha_o \xi_\ell) - E_1(\alpha_o \xi_u)]$$

$$+ \sum_{s=1}^{N} \frac{f_s}{\alpha_s^s} \, [\gamma(s, \alpha_s \xi_u) - \gamma(s, \alpha_s \xi_\ell)] \tag{7}$$

where E_1 is the first exponential integral and $\gamma(s,y)$ is the incomplete gamma function. The first term in the series becomes dominant at high energies and leads to the familiar $E^{-1}\ln E$ dependence usually associated with the Born approximation.

The Born approximation breaks down at low energies owing to the distortion of the incoming and outgoing plane waves in the field of the atom or molecule and other physical effects. To account for all of the effects would be a major computational task, losing thereby the simplicity and utility of the Born results. Green and Dutta developed a methodology to embody these distortion effects semi-empirically yet at the same time to preserve the simple feature of their analytic form of the Born approximation. In consideration of experimental data, and a diffraction-like pattern in low-energy angular distributions, they proposed a 'distorted' generalized oscillator strength (DGOS) of the form

$$f(x,E) = [1 - (W/E)]^\tau \, f_B(\xi) + C[1 - (W/E)]^\delta \, (W/E)^m \, \xi \exp{-\gamma\xi} \tag{8}$$

where τ, C, δ, m and γ are adjustable parameters and $f_B(\xi)$ is the analytic representation of the Born result in the form of Eq. (6).

Figure 1 illustrates a set of analytic DGOS inferred by Jusick et al.[24] for e-He impact leading to excitation of the 2^1P state. Such a DGOS implies the differential inelastic cross section for scattering into any angle (θ) at any incident energy (E). The DGOS can also be analytically integrated to give the inelastic cross section (integrated over all angles of scattering) at any incident energy (E) for the specific energy loss or excitation energy W.

The techniques applied to discrete excitations can be used with some generalizations for the ionization continuum. Now energy loss is a continuous variable, $W = I + T$ where T is the energy of the secondary electrons and I is the ionization threshold. The GOS calculation is similar to that leading to Eq. (3) except that now I is the transition matrix element connecting the ground state to the continuum wave function for the outgoing electron.

80

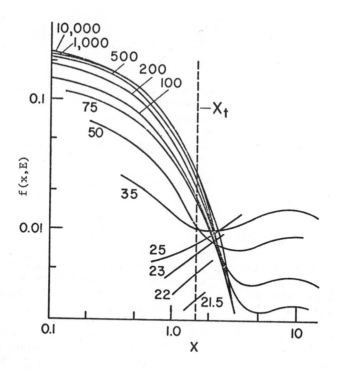

Fig. 1. Illustrating the GOS and DGOS for the 1s-2p excitation of He.
The curve for E = 10,000 eV may be viewed as the BBA GOS.
All lower energy curves represent DGOS's inferred from experi-
ment and parametrized using Eq. (8) (Ref. 24).

Massey and Mohr[25] (MM), using hydrogenic continuum wavefunctions
along with the Born-Bethe approximations, carried out the earliest
detailed theoretical calculation of an ionization continuum GOS.
They presented a contour plot of their results as functions of x and
the variables w = W/I. The MMB surface F(x,w) has a peak located
near $x \simeq 0$, $w \simeq 1$, which falls off and becomes less steep as one
descends along a ridge $x \approx w$. The approach of Green and Dutta was
to represent the MMB surface by an approximate analytic representation. Their formula facilitated distorting and scaling the MMB
surface into a more general surface which satisfies available experimental and theoretical constraints including optical oscillator
strengths.

Green and Dutta attempted to correct for the low energy failure
of the BBA by using a distorted continuum generalized oscillator
strength (DGOS) in a fashion similar to that used for excitation
GOS. These semi-empirical techniques in conjunction with extensive
compilations of experimental data have been used as the basis for
inferring reasonably complete sets of excitation and ionization
cross sections for the gases He[24], O[26], N_2[27], O_2[28], CO_2[29], CO[30],
H_2[31], H_2O[32,33] and Ar[34]. These representations have been recently
updated by Jackman et al.[35]

III. INDEPENDENT PARTICLE MODEL (IPM)

Whereas the initial UF studies leaned on hydrogen or helium wave functions, more recent UF studies use wave functions obtained from numerical solutions of the radial Schrödinger equation for realistic electron-atom potentials. The Schrödinger equation may be written as

$$\left[\frac{d^2}{dr^2} - \frac{\ell(\ell + 1)}{r^2} - V(r) + E \right] P(r) = 0 \qquad (9)$$

where r is in units of Bohr radii (a), and energies are in Rydbergs. Here V(r) is a central atomic potential due to the Z units of nuclear charge and the average effect of the remaining electrons upon the active electron. The development of rapid numerical techniques for solving the radial Schrödinger equation permits one to systematize a large body of physics in terms of realistic electron-atom or electron-ion potentials. At the University of Florida we have used an IPM potential of the Green Sellin Zachor[36] (GSZ) form

$$V(r) = -2r^{-1}[(Z - \eta)\Omega + \eta] \qquad (10)$$

where η is the degree of ionicity and Ω is an analytic screening function given by

$$\Omega(r) = [H(e^{r/d} - 1) + 1]^{-1} \qquad (11)$$

where H is a shape parameter and d is a scale parameter. The ionicity takes on the values $\eta = 1,2,3...$ in a series such as Fe I, Fe II, Fe III... When the parameters H and d are adjusted by the optimizing procedure of Bass et al. [37] the analytic potentials are very close to the average Hartree-Fock potential leading to total energies which are usually only 20-40 ppm above Hartree-Fock.

Other IPM potentials have been used extensively including the Hartree-Fock-Slater potential of Herman and Skillman[38], the Slater Xα potential[39], the optimum numerical potential of Talman et al.[40] and others. Comparative studies by Talman, et al.[41] show that the optimum IPM is the most accurate IPM simulation of Hartree Fock leading to total energies usually only 10-25 ppm higher than Hartree-Fock. Unfortunately, both the analytic optimized and numerical optimized potentials lead to eigenvalues which depart somewhat from experimental single particle excitation energies. For this purpose, empirical adjustment of the GSZ parameters as carried out by Ganas and Green[42] have been used to "tune" the theoretical configuration averages to experiment. Figure 2a illustrates the type of agreement achieved for the excited states of OI[43]. Here the lines denote orbital averages of experimental levels, the symbols ⊡ denote the GSZ-IPM levels for the parameters d = 0.8164 and H = 2.224. Note that the valence state $2p^4$ is much lower at about -1.00 Ry. Figure 2b illustrates Born excitation cross sections calculated with the GOS's obtained from the GSZ-IPM Schrödinger wavefunctions.

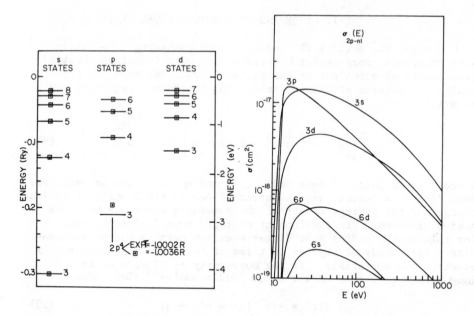

Fig. 2. (a) Excited states of OI; (b) Born excitation cross
sections (Ref. 43).

Berg and Green[44], Kazaks et al.[43] and Bass et al.[45] have calculated the continuum GOS for several gases and used these results
to calculate the differential ionization cross sections, S(E,T).
Figure 3 illustrates some results for Ne and Ar.

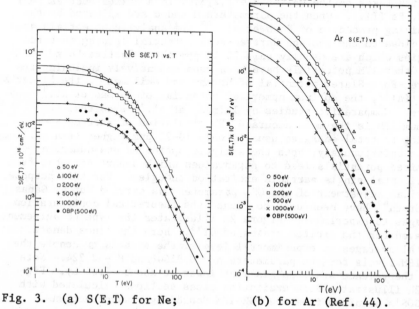

Fig. 3. (a) S(E,T) for Ne;   (b) for Ar (Ref. 44).

The theoretical results shown in Fig. 3 which generally are compatible with experiment[46] imply that no matter what the primary energy, the secondary electrons predominantly have low energies (<50 eV).

It is helpful to represent the differential ionization cross sections extracted from experimental or theoretical GOS data by simple analytic functions of E. Good representations have been achieved for secondary electron distributions using the Lorentz form[47]

$$S(E,T) = A(E) \frac{\Gamma^2}{(T - T_0)^2 + \Gamma^2}$$ (12)

where the parameters A, T_0, and Γ can all be dependent upon the primary energy. For this form the total ionization cross section can then be integrated analytically giving

$$\sigma_i(E) = \int_0^{T_m} S(E,T)dT = A\Gamma[\tan^{-1}[(T_m - T_0)/\Gamma] + \tan^{-1}(T_0/\Gamma)]$$ (13)

where $T_m = (1/2)(E - I)$ is the maximum energy one can assign to the secondary electrons if we use the rule that the faster outgoing electron is the primary. Green and Sawada[47] found reasonable smooth representations of the parameters $T_0(E)$, $\Gamma(E)$, and $A(E)$

$$T_0(E) = T_s - [T_a/(E + T_b)]$$ (14)

$$\Gamma(E) = \Gamma_s[E/(E + \Gamma_b)]$$ (15)

and

$$A(E) = \sigma_0(K/E) \ln (E/J)$$ (16)

where T_s, T_a, T_b, Γ_s, Γ_b, K, and J are adjusted parameters and $\sigma_0 = 10^{-16}$ cm^2. Once the first five parameters are fixed, the last two parameters, K and J, can be evaluated directly from total ionization cross section data. At high energies $\sigma_i(E) \to E^{-1} \ln E$, which is the traditional asymptotic form for the total ionization cross section in the Born-Bethe approximation. The fits with these forms have been very good. The form of S(E,T) can also be used to analytically evaluate the ionization contribution to the loss function as discussed in the next section.

IV. ENERGY DEPOSITION BY THE INTEGRAL CSDA METHOD

Having assembled a DACS it is possible to carry out an energy deposition analysis, in which we determine the energy deposited in various excitations (or equivalently, the number of such excitations) as a result of the complete degradation of an electron with given initial energy. To do this by the CSDA method we first calculate the loss function

$$L(E) = - \frac{1}{n} \frac{dE}{dx} \tag{17}$$

where n is the number density of the gas. The expression for $L(E)$ in terms of all the cross sections is given by

$$L(E) = \sum_j W_j \sigma_j (E) + \int_0^{(E-I)/2} (I+T) \, S(E,T) \, dT \tag{18}$$

where σ_j is the excitation cross section for this state, and I is the ionization threshold. We can then calculate the secondary spectrum

$$n(E,T) = \int_{2T+I}^{E} \frac{S(E',T)}{L(E')} \, dE' \tag{19}$$

where I is the threshold for the ionization continuum. Finally we may calculate the population of the jth excited state [9]

$$J_j(E) = \int_{W_j}^{E} \frac{\sigma_j(E')}{L(E')} \, dE' + \int_0^{(E-I)/2} J_j(T) n(E,T) \, dT \tag{20}$$

Here the first term represents that contribution due to the primary electron alone, while the second term gives the contribution resulting from all higher generations.

Bass et al. carried out an analysis on mercury using published[48-50]   ionization cross sections. The loss functions and eV/ion pair are shown in Figure 4a. The calculated results are seen to be in good agreement with the experimental result of Jesse[51] (□). Efficiencies for production of various excited states ($W_j J_j /E$) and for ionization ($I J_j /E$) are shown in Figure 4b. These results differed insignificantly for the two choices of ionization cross section (S for Smith[48], L for Liska[49] and B for Bleakney[50]).

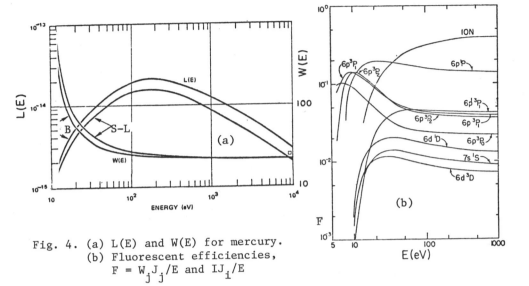

Fig. 4. (a) L(E) and W(E) for mercury.
(b) Fluorescent efficiencies,
$F = W_j J_j / E$ and IJ_i / E

The problem of including energy loss of electrons to thermal electrons in a plasma has been studied by Butler and Buckingham[52] and other authors[53-55]. It is possible to approximately represent such processes by a simplified loss function of the form

$$L = -\frac{1}{N_e}\frac{dE}{dx} = \frac{A}{E}\left(1 - \frac{E_a}{E}\right) \qquad (21)$$

where for present purposes $A = 3.0 \times 10^{-12}$, and $E_a = 10^{-4} T_e$ where T_e is the thermal electron temperature in degrees Kelvin, N_e is the electron density in particles per cc and E is in eV. This is a simplified version of a formula proposed by Swartz et al[56] which should be good in the low energy electron domain and for electron densities in the $10^4 - 10^{12}$ range.

The problem of calculating excitations when the medium is a mixture of constituents has been treated by Stolarski and Green[57] and by others.

The CSDA approach provides a convenient and rapid system of calculating excitation efficiencies. It is known to lead to inaccuracies [9,20] of the order of 10 or 20% in calculating populations or yields. Such inaccuracies are usually of minor consequence as compared to inaccuracies or incompleteness of the atomic cross section sets. Furthermore, such inaccuracies are of no consequence in the early phase of an analysis of a physical idea such as the radioactive excitation model of late time supernova luminosity. However, esthetically one always wants to apply the best physics to a problem if the penalty is not too great. The yield spectra approach appears to be such a method.

V. YIELD SPECTRA

In this section we concern ourselves with a method of energy apportionment which allows for the discrete nature of the individual energy loss mechanisms yet appears to have the simplest possibility for applications. As accurate cross section sets became available at the University of Florida for aeronomical and radiation physics studies it was purposeful to utilize a more accurate energy apportionment methodology. Peterson[10] developed a discrete energy bin method (DEB) in which the energy range between some initial value and the threshold of the state of interest is divided into bins. An idealized degradation process is then assumed to commence in which the initial electron is fractionally redistributed into the lower-energy bins. This idealized process is continued as each energy bin is emptied in turn until all the bins above and including the bin containing the lowest threshold have been emptied. In this way the mean total number of excitations of each state produced in the complete degradation of an electron from a given incident energy is obtained.

Jura[11], Dalgarno and Lejeune[12], and Cravens et al.[13] modified the DEB method to obtain an equilibrium flux subsequently used to calculate state populations. This equilibrium flux $f(E,E_o)$ is equivalent to the degradation spectrum of Spencer and Fano[4,5,15]. The yield spectra approach[16-18] is a further modification of the equilibrium flux method developed in consideration of applications. Thus, the yield spectrum as a basic distribution function, is much simpler than the degradation spectrum or equilibrium flux. Hence its physical implications are more transparent and it is more amenable to convenient analytic representation. The yield spectra may be calculated from the discrete energy bin method using

$$U(E,E_o) = \sigma_\tau(E)f(E,E_o) = N(E)/\Delta E \qquad (22)$$

Here $\sigma_\tau(E)$ is the total inelastic cross section, and $N(E)$ is the number of electrons in the bin centered at E after one bin has been emptied and before the next lower nonempty bin of width ΔE centered at E is considered. By utilizing the DEB method in this mode rather than that of Peterson[10] it was also possible to use wider bin widths[18] which greatly reduces the time and cost of the DEB method permitting extension of the method to high energies. We will refer to this use of the DEB to obtain yield spectra, as the modified discrete energy bin method (MDEB).

To calculate the number of excitations of a state j associated with the complete degradation of a primary electron of energy E_o we use

$$J_j(E_o) = \int_{W_j}^{E_o} U(E,E_o)p_j(E)dE \qquad (23)$$

where $p_j(E) = \sigma_j(E)/\sigma_\tau(E)$ is the probability for excitation of the jth state with excitation energy W_j. This equation is equivalent to the corresponding equation for populations in terms of the degradation spectrum and the cross sections $\sigma_j(E)$. However, the advantages of working with yield spectra and probability of excitation rather than degradation spectra and cross sections are quite substantial since, except at very low energies, $U(E,E_0)$ and $p_j(E)$ both vary with E and with substance in a much simpler manner than do $f(E,E_0)$ and $\sigma_j(E)$. Hence the numerical evaluation of (22) is more efficient than the corresponding equation based upon degradation spectra. Indeed, at high energies $U(E,E_0)$ becomes very flat as does $p_j(E)$ for allowed states of excitation. Thus, from the gross form of (22) we would expect the integral to approach a constant times E_0. Hence, a specific yield $J_j(E_0)/E_0$ should approach a constant at higher primary energies. The same remarks apply to specific yields for ionization and dissociation. Since for forbidden states the probability rises and then falls rapidly at higher energies, the resulting yields are somewhat more complicated than for allowed states.

Because of the simple nature of the $U(E,E_0)$ it is natural to represent it analytically, thus continuing a philosophy used at the University of Florida[7,22,57] in their CSDA methods. Here, however, we are allowing for the discrete nature of the slowing down process. The analytic representations of $U(E,E_0)$ permit us to infer important excitation properties of gases with a degree of accuracy which should suffice for many applications. If greater accuracy is needed the numerical yield spectra can be employed.

Using cross sections such as those described in Section II, we have calculated yield spectra for nine atmospheric gases, for a range of incident energies E_0 from 350 eV to 10 keV. We may represent the yield spectra by

$$U(E,E_0) = U_a(E,E_0)\theta(E_0 - E - E_\theta) + \delta(E_0 - E) \qquad (24)$$

Here, θ is the Heaviside function with E_θ, the minimum threshold of the states considered, $\delta(E_0 - E)$ is the Dirac delta function which allows for the contribution of the source itself, and $U_a(E,E_0)$ can be approximately represented by

$$U(E,E_k) = C_0 + C_1 X + C_2 X^2 \qquad (25)$$

where C_0, C_1 and C_2 are external parameters and $X = E_k^{\Omega}/(E+L)$ where E_k is the incident electron energy in keV and where $\Omega = 0.585$ and $L = 1eV$ are intrinsic parameters. This analytic representation is somewhat more convenient than our original parametrization. Our major reason for adopting this new representation is the evidence (Green et al.,[17] Jackman et al.[58]) that the first two terms of Eq. (25) mainly represent the primary yield spectra whereas $C_2 X^2$ represents the yield spectra due to secondaries and all higher generations. Figure 5a illustrates the analytic yield spectra for several atomic gases. Figure 5b shows the yield spectra for H_2[18] up to 10 MeV indicating that the simple behavior carries over to the relativistic domain. Fig. 5c shows the probabilities of excitation to allowed P_B, forbidden P_b and ionization states.

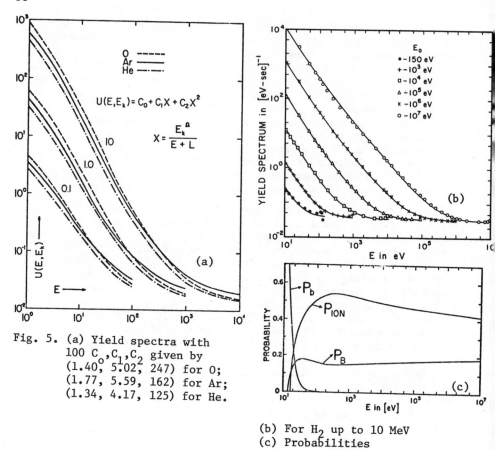

Fig. 5. (a) Yield spectra with
100 C_0, C_1, C_2 given by
(1.40, 5.02, 247) for O;
(1.77, 5.59, 162) for Ar;
(1.34, 4.17, 125) for He.

(b) For H_2 up to 10 MeV
(c) Probabilities

VI. SPATIAL YIELD SPECTRA

The yield spectrum (YS) concept provides a simple two dimensional distribution function which describes the energetic aspects
of the entire electron degradation process. More recently, Jackman
and Green[59], using a Monte Carlo (MC) technique calculated numerical
"spatial-yield" spectra for electron energy degradation in molecular
nitrogen gas. To speed up the low energy component of the calculation they used the results of an elastic multiple scattering calculation of Kutcher and Green[60]. Jackman and Green consider mono-
directional, monoenergetic electrons covering an energy range from
0.1 to 5 KeV impinging initially along the positive z-axis upon an
infinite medium. They obtain numerical longitudinal and radial
(spatial) yield spectra resulting from the entire electron degradation process which can be employed to calculate a yield for any
inelastic state at any position in the medium. Green and Singhal[61]
subsequently found a tractable model for analytically representing
the four dimensional yield spectra (4DYS). From the physical nature
of the system, the scattering cross sections at high and low energies

and the production of secondary, tertiary etc. electrons they sur-
mise that the spatial pattern would have a plume-like character such
as a smokestack plume or a rocket plume. Guided by such analogies
they developed a model of the 4DYS based upon three component
"microplumes." Each microplume is characterized by a few parameters
for its forward component and a few for its backward component and
one or two energy dependent amplitude parameters and a three parameter energy dependent scale factor.

The four dimensional yield spectrum (4DYS) results after an
incident electron of energy E_O and all its secondaries, tertiaries,
etc. have been completely degraded in energy. We define it (in
units of $\#/eV/(gm/cm^2)^3$) by

$$U(E,r,z,E_o) = \frac{N(E,r,z)}{\Delta E \ \Delta z \ \Delta S} \tag{26}$$

$$\Delta P = \pi \left[(r + \frac{\Delta r}{2})^2 - (r - \frac{\Delta r}{2})^2 \right] \tag{27}$$

where $N(E,r,z)$ is the total number of inelastic collisions that
exists in the spatial interval $\Delta z \ \Delta S$ around (r,z) and in the energy
interval ΔE centered at E. Here z is the longitudinal distance
along z-axis and $r = \sqrt{x^2+y^2}$ is the radial distance, both scaled by
an "effective range" $R(E_o)$ (in gr/cm^2).

Equation (26) without the spatial degrees of freedom defines
the original two dimension yield spectra (2DYS) studied by Green
et al.[16-18]. Equation (25) serves as an integral constraint upon
the analytic microplume model for the four variable yield spectra
generated by the Monte Carlo method. After considerable experi-
mentation Green and Singhal selected a model of the form

$$U_a(E,E_k,r,z) = \sum_{i=0}^{2} \frac{A_i}{R^3} \ X^i \ G_i(r,z) \tag{28}$$

where each G_i is a microplume of the form

$$G_i(r,z) = \exp - \left(\frac{\alpha_i r}{1+\delta_i z} + \beta_i^2 z^2 - \gamma_i z \right) \tag{29}$$

and

$$R = R_o \ E_k^q \ e^{\tau/E_k} \tag{30}$$

In terms of the 4DYS the spatial yield of any state may be found using

$$J_j(r,z,E_o) = \int_{W_j}^{E_o} \rho^3 \ p_j(E)U(E,r,z,E_o)dE \quad (in \ \#cm^{-3}) \tag{31}$$

where ρ is the density in gr/cc and $p_j(E) = \sigma_j(E)/\sigma_T(E)$ is the probability of exciting the jth state. Since these do not change the component $G(r,z)$ functions but only the weighting of these functions it is simple to generate contour maps for the various processes.

Singhal et al.[62] have extended and refined the 4DYS of Green and Singhal. Their model of the 4DYS is essentially the same except that now a few of the microplume shape parameters have explicit energy dependence. The 4DYS microplume model has been further generalized to incorporate the fifth degree of freedom, i.e., the polar angle of the electron at any position.

The microplume model implies that each composite process plume has its own intrinsic shape and energy dependence. These extra complexities differ from the results of previous Monte Carlo calculations of Berger et al.[63], and also models by others based upon phenomenological representations of experimental emission contours (Grün,[64]) in which all processes are assumed to have the same spatial dependence.

VII. APPLICATIONS TO LATE TIME SN I SPECTRA

The problem of building a bridge between excitation-deposition theory developed for aeronomy and radiation physics (see Sections I-VI) and observations of late time supernova spectra is formidable indeed. The best route probably is to establish connections to the path pioneered by Meyerott[65] in his radioactive excitation source model and to attempt to improve upon his approximations. Meyerott has calculated the synthetic spectra of Fe^+ and Fe^{++} as a function of electron temperature and obtains results which compare favorably with the observed spectra of SN I 1972e in NGC 5253, taken 245 days after maximum. He uses an electron temperature of $\sim 4000°$ K and the ratio of Fe^{++} to Fe^+ of ~ 4 to 1 which he derives from the ionization by primary electrons with energies of ~ 1 MeV. The secondary electrons produced in the ionization process are assumed to deposit their energy in the electron gas whose temperature $\sim 4000°$ K, is determined by the rate of energy deposited in the electron gas and rate of radiation of forbidden line radiation by Fe II and Fe III. Meyerott finds that approximately 60% of the energy deposited should emerge in forbidden line radiation and the remaining energy should be mostly recombination radiation. His model is consistent with the interpretation that the SN envelope at late time consists mostly of Fe, that 0.5 M⊙ of Fe radiates the observed luminosity and that the same mass of explosively synthesized Ni is required to produce the excitation source.

Clearly, the energy deposition problem as it relates to supernova spectra has a number of unique aspects that have not been addressed in the aeronomical and radiation physics literature. Of particular importance is the need to extend studies based on neutral media or lightly ionized gases to mixtures including highly ionized gases such as Fe II, Fe III, Co II, Co III etc., as well as the electron gas.

Since deposition parameters for such species can not be determined
experimentally, it will be necessary to determine the important
atomic properties by theory.

In applying atomic theory to this supernova problem it will undoubtedly be necessary to strike a compromise between the most
rigorous types of atomic calculations and the crudest types of calculation which are usually based upon hydrogenic wavefunctions. The
former would be too time consuming, whereas the latter would probably
be too inaccurate. In this regard, atomic independent particle
models would seem to commend themselves as reasonable compromises to
derive deposition quantities for the various atomic and ionic com-
ponents of the time varying SN mixtures. For example, Fig. 6 presents IPM energy levels (dots) for Fe I, Fe II and Fe III from the
analytic GSZ potential with parameters (H,d) (2.936, 0.7645); (2.771,
0.6901); and (2.620, 0.6265) respectively obtained by minimization
of the Hartree-Fock total energies[66]. Also shown by the pluses are
the Hartree-Fock eigenvalues. It is seen that the IPM provides a
reasonable system of inner levels which carries over to the excited
state levels.

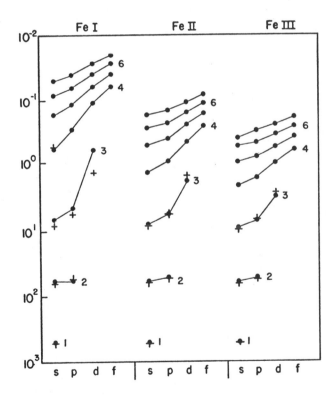

Fig. 6. Hartree-Fock and IPM eigenvalues for Fe I, Fe II and Fe III.
The energies are in Rydbergs.

The wavefunctions from this IPM may be used to obtain GOS for both discrete states and the continuum. From these GOS one can generate excitation and ionization cross sections which can be used to determine the quantities needed in the CSDA method of Section IV or the yield spectra method of Section V. It would be interesting to see, for example, whether the rules of Meyerott (his Eqs. 2-4) are confirmed by this calculation.

The decomposition of excited state populations calculated in this way can be disaggregated in a reasonable way if multiplet splittings are known from optical oscillator strengths. Clearly there is a need to connect these simple atomic IPM models semi-empirically to the more detailed spectroscopic models which are needed to give reasonably accurate spectral line positions. In addition, there is a further need to relate collisional quantities as generalized oscillator strengths to spectroscopic quantities such as optical oscillator strengths.

As to energy deposition techniques themselves, there is a need to develop methods which can be used with realistic atomic and ionic mixtures for the beta and gamma energy ranges of interest. This will undoubtedly require that those working on the energy deposition problem stay in good communication with SN observers and modelers and with the atomic physicists calculating atomic properties by more rigorous methods.

Finally we should note that a number of fundamental questions remain to be answered. For example, the yield spectral distributions which we use to calculate populations are quite different from the Boltzmann distribution used by Meyerott. It is also interesting to note that the large low energy peak which we find for the probability for exciting forbidden states (see P_b in Fig. 5c) favors the idea that forbidden line radiation arises from secondary electrons. However, we have not yet incorporated the electron gas into the yield spectra method so that the detailed equivalence is not yet established. Clearly much work remains.

ACKNOWLEDGMENTS

This research was supported by the Planetary Atmospheres program of the National Aeronautics and Space Administration under grant NGL-10-005-008 and the Radiological Physics Program of the Department of Energy. The author would also like to thank Dr. J. R. Buchler, M. Livio and R. E. Meyerott for introducing him to the literature of supernova, Dr. P. S. Ganas for his help with Figure 6, and Mr. John Merts for his careful reading of the manuscript.

REFERENCES

1. N. Bohr, Phil. Mag. 25, 10, (1913; 30, 581 (1915).
2. H. Bethe, Ann. Phys. 5, 325 (1930).
3. R. H. Fowler, Proc. Camb. Phil. Soc. 21, 531 (1922).
4. L. V. Spencer and U. Fano, Phys. Rev. 93, 1172 (1954).
5. U. Fano., and L. V. Spencer, Int. J. Rad. Phys. Chem. 7, 63 (1975).
6. M. J. Berger, Methods of Computational Physics (New York: Academic Press) (1963).
7. A. E. S. Green and C. A. Barth, J. Geophys. Res. 70, 1083 (1965).
8. S. P. Khare, J. Phys. B: Atom. Molec. Phys. 3, 971 (1970).
9. L. R. Peterson and A. E. S. Green, J. Phys. B, 1, 1131 (1968).
10. L. R. Peterson, Phys. Rev. 187, 105 (1969).
11. M. Jura, PhD Thesis, Harvard University, (1971).
12. A. Dalgarno and C. Lejeune, Planet. Space Sci. 19, 1653 (1971).
13. T. E. Cravens, G. A. Victor and A. Dalgarno, Planet. Space Sci. 23, 1059 (1975).
14. D. A. Douthat, Radiat. Res. 62, 1 (1975); 64, 141 (1975).
15. U. Fano, Radiat. Res. 64, 217 (1975).
16. A. E. S. Green, C. H. Jackman and R. H. Garvey, J. of Geophys. Res. 82, 5104 (1977).
17. A. E. S. Green, R. H. Garvey and C. H. Jackman, International J. of Quan. Chem. 97, 103 (1977).
18. R. H. Garvey, H. S. Porter and A. E. S. Green, J. of Applied Physics, 48, 4353 (1977).
19. A. R. P. Rau, M. Inokuti, and D. A. Douthat, Phys. Rev. A, 18, 971 (1978).
20. R. H. Garvey and A. E. S. Green, Phys. Rev. 14, 946 (1976).
21. E. N. Lassettre, et al., J. Chem. Phys., 40 1208, 1218, 1222, 1232, 1242, 1248, 1256, 1271, (1964).
22. A. E. S. Green and S. K. Dutta, J. Geophys. Res. 72, 3922 (1967).
23. N. F. Mott and H. S. W. Massey, The Theory of Atomic Collisions, Clarendon Press, London (1965).
24. A. T. Jusick, C. E. Watson, L. R. Peterson and A. E. S. Green, J. Geophys. Res. 72, 3943 (1967).
25. H. S. W. Massey and C. B. Mohr, Proc. Roy. Soc. (London) A, 132, 613 (1933).
26. A. E. S. Green and C. A. Barth, J. Geophys. Res. 72, 3975 (1967).
27. R. S. Stolarski, V. A. Dulock Jr., C. E. Watson and A. E. S. Green, J. Geophys. Res. 72, 3953 (1967).
28. C. E. Watson, V. A. Dulock Jr., R. S. Stolarski and A. E. S. Green, J. Geophys. 72, 3961 (1967).
29. D. J. Strickland and A. E. S. Green, J. Geophys. Res. 74, 6415 (1969).
30. T. Sawada, D. L. Sellin and A. E. S. Green, J. of Geophys. Res. 77, 4819 (1972).
31. W. T. Miles, R. Thompson and A. E. S. Green, J. of Applied Physics, 43, 678 (1972).
32. A. E. S. Green, J. J. Olivero and R. W. Stagat, in: Biophysical aspects of radiation quality (IAEA, Vienna, 1971) p. 79.
33. J. J. Olivero, R. W. Stagat and A. E. S. Green, J. Geophys. Res. 77, 4797 (1972).

34. L. R. Peterson and J. E. Allen Jr., J. Chem. Phys. 56, 6068 (1972).
35. C. H. Jackman, R. H. Garvey and A. E. S. Green, J. Geophys. Res. 82, 5081 (1977).
36. A. E. S. Green, D. L. Sellin and A. S. Zachor, Phys. Rev. 184, 1 (1969).
37. J. N. Bass, A. E. S. Green and J. H. Wood, Adv. Quant. Chem. 7, 263 (1973).
38. F. Herman and S. Skillman, Atomic Structure Calculations (Prentice-Hall, Englewood Cliffs, N.J., 1963).
39. K. Schwarz, Phys. Rev. B5, 1355 (1972).
40. J. D. Talman and W. F. Shadwick, Phys. Rev. A14, 36 (1976).
41. J. D. Talman, P. S. Ganas and A. E. S. Green, International J. of Quan. Chem. (in press).
42. P. S. Ganas and A. E. S. Green, Phys. Rev. A4, 182 (1971).
43. P. A. Kazaks, P. S. Ganas and A. E. S. Green, Phys. Rev. A6, 2169 (1972).
44. R. A. Berg and A. E. S. Green, Adv. Quant. Chem. 7, 277 (1973).
45. J. N. Bass, R. A. Berg and A. E. S. Green, J. Phys. B, 7, 1853 (1974).
46. C. B. Opal, E. C. Beaty and W. K. Peterson, At. Data 4, 209 (1972).
47. A. E. S. Green and T. Sawada, J. Atmos. Terr. Phys. 34, 1719 (1972).
48. P. T. Smith, Phys. Rev. 37, 808 (1931).
49. J. W. Liska, Phys. Rev. 46, 169 (1934).
50. W. Bleakney, Phys. Rev., 35, 139 (1930).
51. W. P. Jesse, J. Chem. Phys. 55, 3603 (1971).
52. S. T. Butler and M. J. Buckingham, Phys. Rev. 126, 1-4 (1962).
53. Y. Itikawa and O. Aono, Phys. Fluids 9, 1259 (1966).
54. R. W. Schunk, P. B. Hays and Y. Itikawa, Planet. Space Sci. 19, 125 (1971).
55. K. Takayanagi, and Y. Itikawa, Space Sci. Rev. 11, 380 (1970).
56. W. E. Swartz and J. S. Nisbet, J. Geophys. Res. 76, 8425 (1971).
57. R. S. Stolarski and A. E. S. Green, J. Geophys. Res. 72, 3967 (1967).
58. C. H. Jackman, R. H. Garvey and A. E. S. Green, J. Phys. B. Atom. Molec. Phys. 10, 2873 (1977).
59. C. H. Jackman and A. E. S. Green, J. Geophys. Res. 84, 2715 (1979).
60. G. J. Kutcher and A. E. S. Green, J. Appl. Phys. 47, 2175 (1976).
61. A. E. S. Green and R. J. Singhal, Geophys. Res. Lett. 6, 625 (1979).
62. R. J. Singhal, C. H. Jackman and A. E. S. Green, J. Geophys. Res., 1980, (in press).
63. M. J. Berger, S. M. Seltzer and K. Maeda, J. Atmos. Terr. Phys. 32, 1015 (1970); 36, 591, (1974).
64. A. E. Grün, Z. Naturforschg. 12a, 89 (1957).
65. R. E. Meyerott, Chap. , this volume.
66. R. H. Garvey, C. H. Jackman and A. E. S. Green, Phys. Rev. A 12, 1144 (1975).

CHARGE AND ENERGY TRANSFER IN HEAVY PARTICLE COLLISIONS

R. E. Olson

Molecular Physics Laboratory, SRI International
Menlo Park, CA 94025

ABSTRACT

A short discussion of the general classes of excitation and charge transfer reactions is presented. Specific emphasis is placed on collision mechanisms applicable to low energy (E ≲ 10 eV) scattering between ions and atoms and the resulting scaling of the cross sections on parameters such as collision energy and ion charge state.

INTRODUCTION

In the interpretation of supernovae spectra, heavy particle collisions greatly influence the atomic and ionic composition of the plasma. In particular, charge and energy transfer collisions modify the time dependence of the predicted spectra and are important in the determination of the charge and energy balance of the supernovae plasmas.[1]

In this paper, we will be concerned with presenting a summary of the scaling dependences of the cross sections which are relevant to low energy inelastic collisions. Excitation transfer and charge transfer reactions involving singly and multiply-charged ions will be discussed.

EXCITATION TRANSFER

In general, the excitation transfer reaction

$$A^* + B \rightarrow A + B^* + \Delta E \tag{1}$$

occurs with high probabilities and large cross sections when the potential curves of the initial and final states lie close to one another (i.e., ΔE is small).

In order to determine rough limits on the magnitude of ΔE which allow large cross sections, we need to refer to the basic collision mechanism which was first described by Stückelberg[2,3] and later extended by Demkov.[4] These authors showed that transitions preferentially occur at an internuclear separation R_c where the coupling matrix element $H_{12}(R)$ is equal to one-half the separation between the diabatic potentials $V_{11}(R)$ and $V_{22}(R)$ which describe the initial and final states:

ISSN:0094-243X/80/630095-07$1.50 Copyright 1980 American Institute of Physics

$$H_{12}(R) = \tfrac{1}{2}\left|V_{11}(R) - V_{22}(R)\right| = \tfrac{1}{2}\Delta V(R). \tag{2}$$

For rough calculations, it is appropriate to substitute ΔE for $\Delta V(R)$ in Eq. (2). The coupling matrix element, in turn, is generally exponential in form because it is primarily determined by the overlap of the initial and final molecular wavefunctions at large internuclear separations:

$$H_{12}(R) = A \exp(-\lambda R) . \tag{3}$$

Numerical calculations then showed the cross section determined within the framework of the Stückelberg-Demkov model can be conveniently, and accurately, presented in terms of reduced parameters[5]

$$Q^* = Q/(\tfrac{1}{2}\pi R_c^{\,2}) \tag{4}$$

and

$$v^* = \frac{2\hbar\lambda v \left(1-\dfrac{V_{11}(R_c)}{E}\right)^{\frac{1}{2}}}{\pi \Delta V(R_c)} \tag{5}$$

In Eq. (5), v is the initial relative velocity of the collision. The reduced cross section is shown in Fig. 1 and displays a maximum at $v^* \approx 3.2$ and a threshold at $v^* \approx 0.5$.

To obtain an estimate of the magnitude of ΔE in (1) which allows rapid transfer of electronic excitation, we will use 1×10^6 cm/sec as a representative low energy collision velocity. In turn, λ in Eq. (3) is approximately equal to the square root of the average ionization potential of A^* and B^* in (1) in units of Rydbergs (1 Rydberg = 13.6 eV). Assuming $\lambda = 1$, we then find the excitation transfer cross section is negligible for $\Delta E \gtrsim 0.15$ eV. Hence, to first-order, it is unreasonable to expect a large excitation transfer cross section if ΔE in (1) is much greater than a few tenths of an eV.

CHARGE EXCHANGE (SINGLY-CHARGED IONS)

Charge exchange involving singly-charged ions

$$A^+ + B \rightarrow A + B^+ + \Delta E, \tag{6}$$

generally behave the same as the excitation transfer case described above. That is, the charge exchange cross section is appreciable

only when ΔE is less than a few tenths of an eV and the velocity dependence of the cross section is similar to that given in Fig. 1.

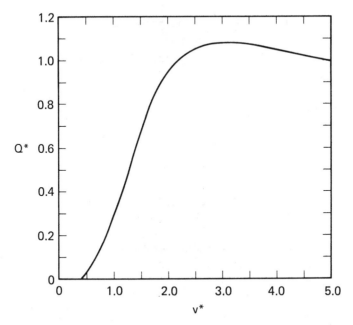

Fig. 1 Reduced cross section plot for near resonant
charge or energy transfer in a two-state system.
The reduced cross section Q^* and the reduced velocity
v^* are given by Eqs. (4) and (5) in the text.

However, an exception to the above behavior is displayed at very
low collision energies, $E \lesssim 0.025$ eV (300°K). In this regime, all
collision trajectories which surmount the centrifugal barrier in the
effective potential of the reactants are accelerated into small internuclear separations where Reaction (6) can proceed with unit probability. A model utilizing the above assumptions has been developed
and is termed the "orbiting" or Langevin model.[6] The basic input to
the model is the long-range intermolecular potential for the reactants in (6), which is a point charge-induced dipole interaction:

$$V(R) = -\frac{q^2 \alpha_d}{2R^4} . \qquad (7)$$

In Eq. (7), α_d is the dipole polarizability of B in atomic units and
q is the charge state of A, which is unity for this example. It then

follows that the charge transfer cross section is given by[6]

$$Q_{q,q-1} = \pi q \left(\frac{2\alpha_d}{E}\right)^{\frac{1}{2}} \tag{8}$$

where all quantities are in atomic units.

As an example, if we assume B in Reaction (6) is atomic hydrogen which has an α_d of 4.50 a_o^3, we find at 0.025 eV or 300°K the charge transfer cross section is given by

$$Q_{q,q-1} = 8.8 \times 10^{-15} \, q - cm^2 \tag{9}$$

Thus, at very low energies charge transfer can proceed with very large cross sections, and in fact may actually increase with decreasing energy as seen by the $E^{-\frac{1}{2}}$ dependence in (8).

CHARGE TRANSFER (MULTIPLY-CHARGED IONS)

Charge transfer involving a multiply-charged ion

$$A^{+q} + B \rightarrow A^{+q-1} + B^+ + \Delta E \tag{10}$$

proceeds via different collision mechanisms than that for the singly-charged case. In reactions involving singly-charged ions, the interaction potentials for both the reactant and product channels are covalent in nature and have similar forms. However, for reactions involving multiply charged ions, the reactant channel is covalent while the product channel is coulombic [i.e., $(q-1)/R$ for Reaction (10)], see Fig. 2. Thus, states which are asymptotically close at large internuclear separations, ΔE small, are well separated at internuclear distances where electron transfer can occur and are not involved in the charge transfer process.

Numerous calculations and experimental measurements confirm that electron transfer preferentially occurs at internuclear separations $R \approx 5$ to 15 a_o (3 to 8 Å) for initial charge states q = +2 to +5. At these distances, the coupling matrix elements are sufficiently large to induce electron transfer from reactant to product channels. If we translate the interaction regions to asymptotic energy separations, we find that charge transfer involving multiply-charged ions proceeds with large cross sections when there are product channels available which are exoergic to the reactant channel by 3 to 10 eV.

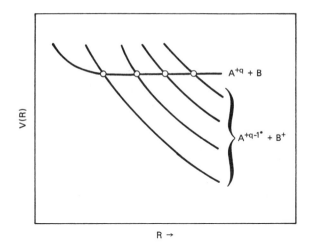

Figure 2 Schematic
of the potential
energy curves for
low velocity charge
transfer between a
multiply-charged
ion and an atom.

There are several general procedures for estimating the magnitude of the charge transfer cross section when there are a large
number of product channels available which cross the initial covalent state in the $R = 5$ to 15 a_o region. One is based on a generalization of the Landau-Zener curve crossing model[7,8] to the multichannel system.[9] Another is based on a tunneling model to describe
the electron transfer.[10] Both models predict the charge transfer
cross section is large, $Q_{q,q-1} \gtrsim 10^{-15}$ cm^2, and is relatively independent of collision velocity until the orbiting region is reached.
Experimental data confirms the models are reasonable, Fig. 3.

If only a few product channels are available for reaction, the
models described above are inapplicable. The charge transfer cross
sections are then highly velocity dependent and one must rely on
difficult experimental measurements or very laborious theoretical
calculations. The calculational procedure is to first calculate the
intermolecular potential curves and coupling matrix elements, and
then solve a set of coupled equations to determine the cross sections. The procedure is very difficult and because of approximations made in determining the potentials, cross sections generally
have error limits of at least ± 25%.

The reverse reaction to (10), charge transfer between ions,
will have extremely small cross sections. This is due to the fact
that the initial channel is coulombic, and at the low energies considered here, the particles will not have sufficient kinetic energy
to reach the internuclear separation region where electron transfer
is possible.

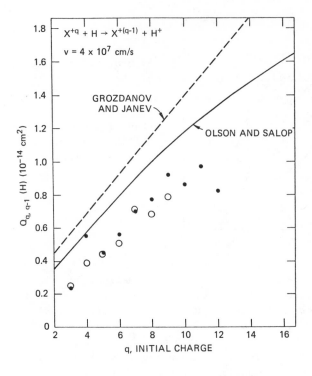

Figure 3 Charge transfer cross sections for ions colliding with atomic hydrogen as a function of initial charge state q at 4×10^7 cm/sec. Open circles are data on Xe^{+q} and solid circles are data on Ar^{+q} measured by Crandall et al. (Ref. 11). Theoretical curves are from references 9 and 10.

SUMMARY

In summary, cross sections larger than 10^{-15} cm^2 can be expected for excitation transfer reactions and charge transfer reactions involving singly charged ions when the asymptotic energy separation ΔE between reactants and products is less than ~ 0.2 eV. For charge transfer involving multiply-charged ions, large cross sections are expected only if product channels exist which are exoergic to the reactant channel by 3 to 10 eV. Charge transfer between two positively charged ions will be negligible at collision energies less than 10 eV.

ACKNOWLEDGEMENTS

Work supported in part by the Aeronomy division of the National Science Foundation.

REFERENCES

1. R. E. Meyerott, Bull. Am. Astron. Soc. 10, 638 (1978).
2. E.C.G. Stückelberg, Helv, Phys. Acta 5, 369 (1932).

3. N. F. Mott and H.S.W. Massey, <u>The Theory of Atomic Collisions</u> (Oxford Press, Great Britain, 1965) pp. 651-4.

4. Yu. N. Demkov, Sov. Phys. JETP <u>18</u>, 138 (1964).

5. R. E. Olson, Phys. Rev. A <u>6</u>, 1822 (1972).

6. G. Gioumousis and D. P. Stevenson, J. Chem. Phys. <u>29</u>, 294 (1958).

7. L. Landau, J. Phys. (USSR) <u>2</u>, 46 (1932).

8. C. Zener, Proc. Roy. Soc. (London) A <u>137</u>, 696 (1933).

9. R. E. Olson and A. Salop, Phys. Rev. A <u>14</u>, 579 (1976).

10. T. P. Grozdanov and R. K. Janev, Phys. Rev. A <u>17</u>, 880 (1978).

11. D. H. Crandall, R. A. Phaneuf, and F. W. Meyer, Phys. Rev. A (in press).

ENERGY LEVELS, WAVELENGTHS AND TRANSITION PROBABILITIES FOR THE FIRST FIVE SPECTRA OF Fe, Co AND Ni

W. L. Wiese
National Bureau of Standards, Washington, D.C. 20234

ABSTRACT

Atomic energy level, wavelength and transition probability data for the first five spectra of Fe, Co and Ni are reviewed, and lists of recent comprehensive data tables are presented. The source material for transition probabilities, both for allowed and forbidden lines, is critically discussed, since these data contain large uncertainties.

INTRODUCTION

The first five spectra of iron, cobalt and nickel originate from atomic structures which possess incomplete shells of 3d electrons. Because of the numerous possible combinations of spin and angular momentum for the outer electrons, including the excited one, a very large number of excited atomic states is obtained and the resultant spectra are very complex. Furthermore, since the outer electrons interact quite strongly, the theoretical treatment of these complicated atomic structures is very difficult and many results have remained unreliable to date. The task of analyzing these spectra precisely and determining their transition probabilities thus has proceeded largely from the experimental side. This, too, has been a very large task and the situation is still in a rather incomplete state, as will be seen below. On the other hand, the progress made in recent years is considerable, and further advances, especially in the area of multiply ionized spectra, may be expected in the not too distant future.

The spectral data base for the fifteen spectra under consideration is rather unevenly distributed. While wavelength data for the neutral spectra of iron, cobalt, and nickel cover essentially the prominent as well as most moderate-strength and weaker observable lines, the data become less complete for the higher stages of ionization up to stage V. The energy level data also are more extensive for the lower spectra, where they include at least some states with principal quantum numbers differing by two or more from the ground state. Furthermore, the data on iron are generally more extensive than those for cobalt and nickel, which is largely due to the strong interest in Fe, both astrophysically and in laboratory plasma applications.

With respect to atomic transition probabilities, the differences in the availability and quality of data from spectrum to spectrum are even more striking. Furthermore, practically all of the transition probability data are less accurate by several orders of magnitude than the wavelength data; they are much less complete and cover only a very small fraction of the identified and classified lines. For most of the multiply charged ions of cobalt and nickel no reliable transition probabilities are available as yet. The decrease in the quantity of data from the first to the fifth spectrum is thus very drastic. Even for the three neutral spectra of Fe, Co and Ni, only for Fe is there a large set of reliable data available.

Wavelength and Energy Level Data

Traditionally, complex spectra have been studied and analyzed by laboratory observations with great precision, and atomic energy levels have been derived from such spectral analyses. The spectra have been usually generated with hollow cathode sources or electrodeless lamps for the case of neutral and singly ionized species and with sliding sparks for the third to fifth stages of ionization, and have been recorded photographically with high resolution spectrographs.[1]

With respect to wavelength data, several comprehensive tables have been assembled, of which the most recent ones are listed below:

a.   Tables of Spectral Line Intensities, W. F. Meggers, C. H. Corliss, B. F. Scribner, NBS Monograph 145, Parts 1 and 2, U.S. Government Printing Office, Washington, D.C. (1975).
b.   Line Spectra of the Elements, in the CRC Handbook of Chemistry and Physics, 60th Edition, CRC Press (1979).
c.   Atomic Emission Lines in the Near Ultraviolet; Hydrogen through Krypton, R. L. Kelly, NASA Technical Memorandum 80268 (1979) (Sections I and II).
d.   Atomic and Ionic Emission Lines Below 2000 Å--Hydrogen through Krypton, R.-L. Kelly and Palumbo, NRL Report 7599, U.S. Government Printing Office, Washington, D.C. (1973).

The most recent papers containing wavelength data on the spectra under consideration are assembled in Table I and provide a supplement to these general sources.

Experimental studies have also been the basis for new critical data compilations of the atomic energy levels of iron, cobalt and nickel through all stages of ionization. These tabulations, all performed at NBS, either have recently been published or are soon to be published as follows:

a.   Energy Levels of Iron, Fe I through Fe XXVI, by J. Reader and J. Sugar, J. Phys. Chem. Ref. Data 4, 353-440 (1975).
b.   Energy Levels of Cobalt, Co I through Co XXVII, by J. Sugar and C. Corliss, J. Phys. Chem. Ref. Data 9 (1980).
c.   Energy Levels of Nickel, Ni I through XXVIII, by C. Corliss and J. Sugar, J. Phys. Chem. Ref. Data 9 (1980).

The energy levels in the above-listed compilations are accompanied by their quantum mechanical designations. The choice of the designations has sometimes produced problems, since often no simple "pure"

TABLE I. Recent Sources of Wavelength Data.

Iron:

Fe I: H. M. Crosswhite, J. Res. Nat. Bur. Stand. (U.S.) 79A, 17 (1975).

U. Litzen and J. Verges, Phys. Scr. 13, 240 (1976) (infrared lines).

Fe II: S. Johansson, Phys. Scr. 18, 217 (1978).

Fe IV: J. O. Ekberg and B. Edlén, Phys. Scr. 18, 107 (1978).

Fe V: J. O. Ekberg, Phys. Scr. 12, 42 (1975).

Cobalt:

Co II: L. Iglesias, Opt. Pur. Apl. (Spain) 12, 63 (1979).

Co IV: R. Poppe, T. A. M. van Kleef, and A. J. J. Raassen, Physica (Utrecht) 77, 165 (1974).

Co V: A. J. J. Raassen and T. A. M. van Kleef, Physica (Utrecht) 96C, 367 (1979).

Nickel:

Ni IV: R. Poppe, Physica (Utrecht) 75, 341 (1974); 81C, 351 (1976).

Ni V: A. J. J. Raassen, T. A. M. van Kleef, and B. C. Metsch, Physica (Utrecht) 84C, 133 (1976).

A. J. J. Raassen and T. A. M. van Kleef, Physica (Utrecht) 85C, 180 (1977).

designation in terms of a single quantum state is appropriate.
Calculated percentage compositions of the levels in terms of the
two largest basis states of a configuration are therefore given,
whenever available; this should be a useful guide to the true
quantum character of the level. All these compilations cover
essentially only atomic states involving principal quantum numbers
which differ from those of the ground state by no more than two.
More highly excited energy levels have usually not been derived
experimentally. An exception is Ni II, where many energy levels of
principal quantum numbers through nine have been determined by
Shenstone.[2] With respect to the 1975 energy level tables on Fe, a
significant addition is the new work by Ekberg and Edlén on Fe IV,
cited in Table I.

Atomic Transition Probabilities (or f-values)

For elements of the iron group, the following critical evalua-
tions and compilations of atomic transition probability data have
been underway at NBS for several years and have been essentially
completed now:

a. Forbidden Lines of the Iron Group Elements, M. W. Smith
and W. L. Wiese, J. Phys. Chem. Ref. Data 2, 85-120 (1973).
b. Allowed Lines of Scandium and Titanium, W. L. Wiese and
J. R. Fuhr, J. Phys. Chem. Ref. Data 4, 263-352 (1975).
c. Allowed Lines of Vanadium, Chromium and Manganese,
S. M. Younger, J. R. Fuhr, G. A. Martin, and W. L. Wiese, J. Phys.
Chem. Ref. Data 7, 495-629 (1978).
d. Allowed Lines of Iron, Cobalt and Nickel, J. R. Fuhr,
G. A. Martin, W. L. Wiese, and S. M. Younger, J. Phys. Chem. Ref.
Data 9 (1980).

Furthermore, a comprehensive NBS data compilation containing all
the material for the iron group, with the earlier material revised
and updated, is planned for early 1981.
In view of the large uncertainties in atomic transition probabilities, a brief discussion of the source material is in order.
It is useful to treat the data for electric dipole (allowed) and
forbidden transitions separately:

(a) Electric dipole transitions. A variety of experimental
and theoretical approaches have been used to determine the data.
For the spectra under discussion, the existing experimental data
are--at this stage--much more accurate than the calculated ones,
and thus constitute the bulk of the above-cited NBS critical data
compilation. Table II gives an overview of the source material
available and lists the number of transitions included in the NBS
tables. The four principal experimental techniques which have
found repeated applications are the emission, the anomalous dispersion, the absorption and the atomic lifetime techniques. All
these techniques are well established and have been reviewed in the

literature.[3] The largest number of data has been obtained from
emission experiments which for the spectra in question have been
usually performed on a relative scale. Atomic lifetime data could
be applied to convert these data to an absolute scale. Anomalous
dispersion ("hook") and absorption measurements have usually been
performed on a relative scale, too, and have also been normalized
with lifetime data.

TABLE II. Availability of transition probability data for allowed
lines and their sources. Listed are the approximate number of lines
given in the above-cited NBS data tables and the approaches used to
obtain the data. Also referenced in parentheses are two comprehensive theoretical papers which provide additional low accuracy data
for many thousands of lines. (This work will be discussed later;
see also Figures 4-6.)

Stage of Ionization	Element		
	Fe	Co	Ni
I	1630 - E,A,H,L(S)	208 - H,E,L(S)	280 - H,E,L(S)
II	70 - E,H,L(S)	0 - (S)	53 - E(S)
III	60 - S,L(S)	0 - (S)	19 - S,L(S)
IV	0 - (S)	0 - (S)	0 - (S)
V	0 - (S)	0 - (S)	0 - (S)

E = emission measurement
A = absorption measurement
H = anomalous dispersion (hook) measurement
L = lifetime measurement
S = semiempirical calculation (see Ref. 9, and
    for Fe III-V and Ni III-V, see also Ref. 10)

The critical problem with all these experiments is to obtain good accuracy and reliability. Since the measurements are relatively complex and require not only measurements of emission or absorption line intensities or "hook" distances, but also the determination of excited state populations, several sources of appreciable uncertainties are encountered. In typical cases the measurement accuracy is roughly of the order of 10 to 25%. A very notable exception are the recent high-precision experiments with a sophisticated absorption technique by Blackwell and co-workers[4,5] where accuracies of 0.5% have been achieved on a relative scale. Such measurements, however, require an enormous amount of care and painstaking attention to detail and cannot be quickly extended to large numbers of lines nor to many atomic species. Thus one should not expect major improvements in the uncertainty situation soon. However, even the present situation represents a vast improvement over the status of our knowledge of just ten years ago when the then accepted scale for transition probabilities of Fe I turned out to be in error by as much as a factor of ten for the highly excited lines.

To provide some illustration of the quality of the experimental data, Figs. 1, 2 and 3 provide a few representative comparisons which show a high degree of mutual consistency between independent experimental data sources, but also indicate still appreciable scatter for individual lines. Figure 1 shows a comparison between the two comprehensive emission experiments which have supplied the large majority of reliable laboratory data on neutral iron. These are the stabilized arc measurements by Bridges and Kornblith[6] for more than 500 lines and by May et al.[7] for more than 1,000 lines. It is seen that the agreement between the two sets of data is generally within 25%. There is, however, some marked disagreement between the f-values for the lines of shortest wavelengths. This concerns especially the four lines which are identified by their wavelengths on the graph. A likely explanation is scattered light problems with radiometric standards in this difficult spectral range. [The data by Bridges and Kornblith appear preferable since these authors minimize the problem by application of appropriate filters.] In Fig. 2, the Bridges and Kornblith emission data[6] are compared to the high precision absorption experiments by Blackwell et al.[4,5] When the ratios are plotted against log gf (g is the statistical weight), as done here, a slight systematic trend is apparent. However, most of the lines which have been measured with high precision by Bridges and Kornblith--these are indicated by open circles--scatter around the data of Blackwell et al. on a relative scale within a bandwidth of about ± 10%, which is the total experimental error estimate. Figure 3 shows a comparison of the Bridges and Kornblith emission data with a recent "hook" experiment by Banfield and Huber[8] where the agreement is very impressive, i.e., in practically all cases better than 25%.

Experimental data are, however, available for only the first and second spectra of Fe, Co and Ni. For the higher spectra, the generation of excited ions in reliable, well-defined sources becomes very difficult, and measurements with the traditional

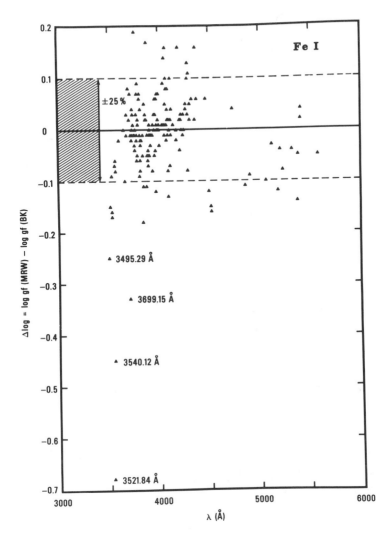

FIG. 1. Comparison of the Fe I emission data of May, Richter, and Wichelmann[7] with those of Bridges and Kornblith.[6] The ratios of gf, i.e. the product of the statistical weight g of the lower atomic state and the absorption oscillator strength f, are plotted versus wavelength.

emission, absorption, or hook methods require large and expensive experimental set-ups. Thus at this time, recourse has to be taken to theory. Several theoretical papers are available in the literature for the third, fourth and fifth spectra of Fe, Co and Ni and two of these--the papers by Kurucz and Peytremann,[9] and Abbott[10]-- are very comprehensive ones, which have yielded data for many

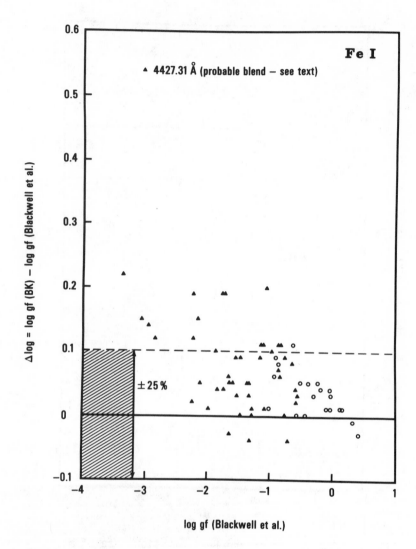

FIG. 2. Comparison of oscillator strengths from the emission measurements of Bridges and Kornblith[6] with the high precision absorption data of Blackwell et al.[4,5] [The ratios are plotted versus the log gf data of Blackwell et al.]

thousands of lines. These papers utilize similar theoretical approaches based on scaled Thomas-Fermi potentials and include limited configuration interaction. Since it is very difficult to assess the accuracy of these calculations for the complex atomic structures encountered, it is very important to check the accuracy of these f-values by comparison with reliable experimental data.

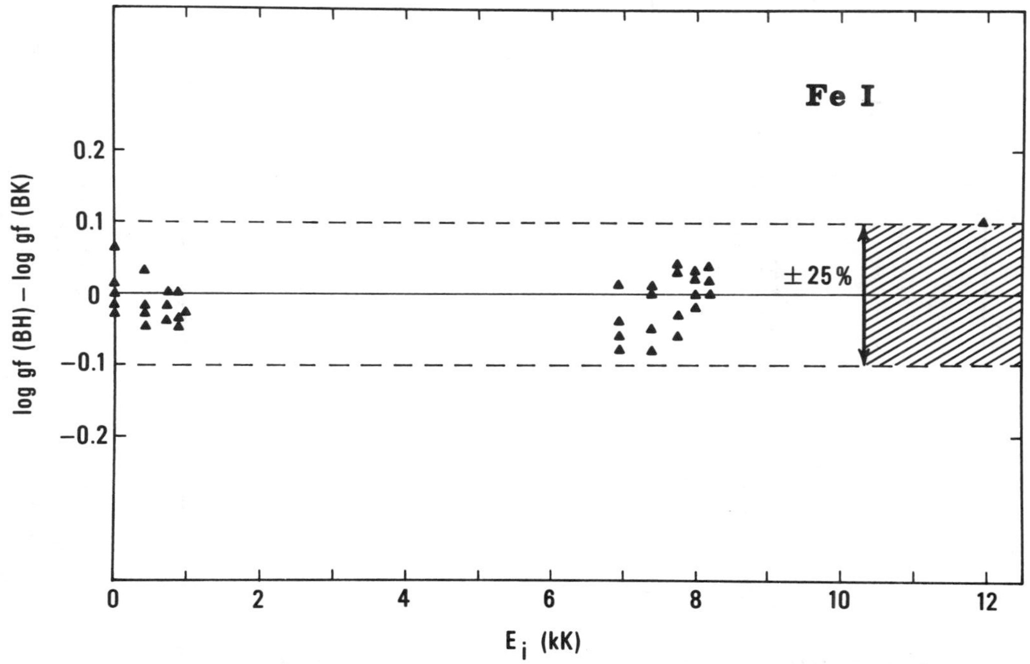

FIG. 3. Comparison of the anomalous dispersion ("hook") data of Banfield and Huber[8] with the emission measurements of Bridges and Kornblith[6] versus lower energy level E_i (in 10^3 cm^{-1}).

For the very extensive calculations of Kurucz and Peytremann--
which include other species of the Fe-group elements--a number of
such comparisons have recently been made[11,12] and several examples
are given in Figs. 4, 5 and 6. Figure 4 shows comparisons with
experimental material for the case of Mn I. This figure is analo-
gous to the earlier graphical comparisons, but the log gf scale is
quite compressed. The shaded area between the solid horizontal
lines is a band for which the agreement between experiment and
calculations is within ± 50%. It can be seen that a substantial

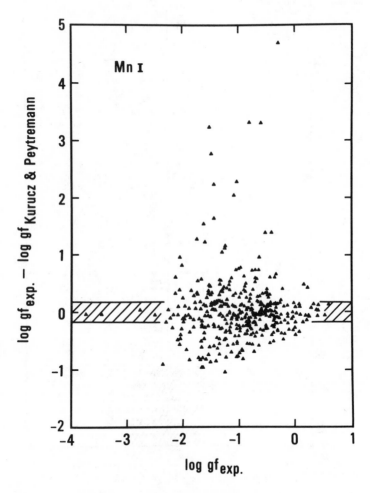

FIG. 4. Comparison of the experimental data[10] for Mn I
with the results of the semi-empirical calculations of
Kurucz and Peytremann.[9] Compared to the earlier figures,
the vertical scale is strongly compressed.

part of the data lie outside this area and some discrepancies are as large as a factor of 10^4. A similar situation is exhibited in Fig. 5 for neutral vanadium. On the other hand, Fig. 6 demonstrates that for a number of resonance lines of neutral iron, agreement with the high precision data of Blackwell et al.[4,5] is generally good; furthermore, the absolute scales appear to be consistent in all cases. Generally, the calculated data of Kurucz and Peytremann seem to agree best with experiment for transitions where the energy levels are only slightly mixed.

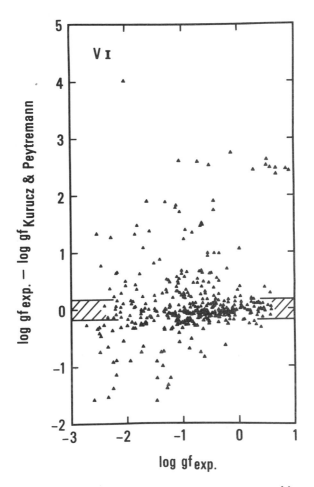

FIG. 5. Comparison of experimental data[10] for V I with the results of the semi-empirical calculations of Kurucz and Peytremann.[9]

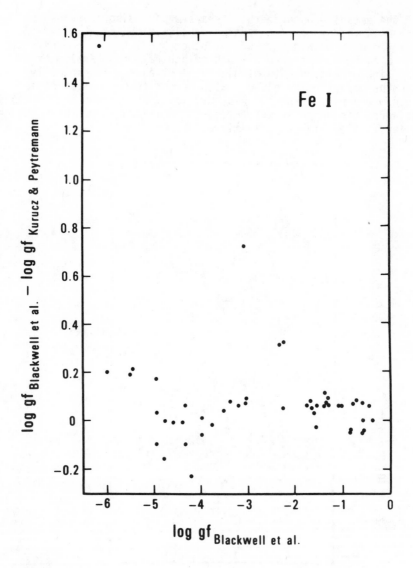

FIG. 6. Comparison of the high precision absorption data of Blackwell et al.[4,5] with the results of the semi-empirical calculations of Kurucz and Peytremann.[9]

(b) Forbidden Lines. "Forbidden lines" are all lines other than electric dipole, i.e., principally magnetic dipole, electric quadrupole and magnetic quadrupole transitions. All material on forbidden lines of the iron-group elements comes from calculations; no experimental data are available. An overview of the availability of data and of their sources is given in Table III. The principal sources for the lower spectra of Fe, Co and Ni are the extensive calculations by Garstang[13-18] and by Grevesse et al.[19] for Fe I. Nearly one-half of these lines have been compiled by Smith and Wiese;[20] these lines account for about 95% of the total strength in each spectrum. For Co II and III some additional material is available[21,22] not compiled in Ref. 20. All calculations involve transitions within and between configurations of the types $3d^n$, $3d^{n-1}$ 4s, and $3d^{n-2}$ $4s^2$, and Garstang[13-18] has repeatedly discussed the method of calculation in detail. Again, as with the theoretical data for allowed lines, the assessment of accuracies for the calculated data is very difficult. The situation is actually worse than for the allowed lines, since essentially no experimental data

TABLE III. Availability of transition probability data for forbidden lines and their sources. The number of lines listed in the NBS data tables[20] and references to the principal literature sources are given. These sources often contain additional data for many weak lines.

Stage of Ionization	Element		
	Fe	Co	Ni
I	103 (Ref. 19)	--	36 (Ref. 16)
II	308 (Ref. 13)	-- [Ref. 21]	60 (Ref. 17)
III	31 (Ref. 14)	-- [Ref. 22]	13 (Ref. 18)
IV	28 (Ref. 15)	--	14 (Ref. 19)
V	19 (Ref. 14)	--	--

are available for comparison. A few direct comparisons with experiment exist for atomic systems much simpler than the Fe-group atoms, and indicate that the calculations of prominent forbidden lines of such atomic systems are reliable to within about 25% or better.[23] It would, however, be dangerous to make any extrapolations to the much more complex atomic systems under consideration here. Fortunately, a few somewhat indirect comparisons[23] with observational data for [Fe II], [Fe III] and [Ni II] are available from astrophysical spectra. All these comparisons indicate that within multiplets the agreement between observations and calculations is remarkably good. On this basis a conservative error estimate for the theoretical data should be a factor of two for prominent and moderate strength lines and probably somewhat better for the strongest lines.

REFERENCES

1. K. Heilig and A. Steudel, in Progress in Atomic Spectroscopy, Part A, edited by W. Hanle and H. Kleinpoppen (Plenum Press, New York, 1979), p. 263.

2. A. G. Shenstone, J. Res. Nat. Bur. Stand. (U.S.) 74A, 801 (1970); 75A, 335 (1971).

3. W. L. Wiese, in Progress in Atomic Spectroscopy, Part B, edited by W. Hanle and H. Kleinpoppen (Plenum Press, New York, 1979), p. 1101.

4. D. E. Blackwell, P. A. Ibbetson, A. D. Petford, and M. J. Shallis, Mon. Not. R. Astron. Soc. 186, 633 (1979).

5. D. E. Blackwell, A. D. Petford, and M. J. Shallis, Mon. Not. R. Astron. Soc. 186, 657 (1979).

6. J. M. Bridges and R. L. Kornblith, Astrophys. J. 192, 793 (1974).

7. M. May, J. Richter, and J. Wichelmann, Astron. Astrophys. Suppl. Ser. 18, 405 (1974).

8. F. P. Banfield and M. C. E. Huber, Astrophys. J. 186, 335 (1973).

9. R. L. Kurucz and E. Peytremann, Smithson. Astrophys. Observ. Spec. Rept. 362 (1975).

10. D. C. Abbott, J. Phys. B 11, 3479 (1978).

11. S. M. Younger, J. R. Fuhr, G. A. Martin, and W. L. Wiese, J. Phys. Chem. Ref. Data 7, 495 (1978).

12. P. L. Smith, Mon. Not. R. Astron. Soc. 177, 275 (1976).

13. R. H. Garstang, Mon. Not. R. Astron. Soc. 124, 321 (1962).

14. R. H. Garstang, Mon. Not. R. Astron. Soc. 117, 393 (1957).

15. R. H. Garstang, Mon. Not. R. Astron. Soc. 118, 572 (1958).

16. R. H. Garstang, J. Res. Nat. Bur. Stand., Sec. A68, 61 (1964).

17. R. H. Garstang, Mon. Not. R. Astron. Soc. 118, 234 (1958).

18. R. H. Garstang, Astrophys. Space Sci. 2, 336 (1968).

19. N. Grevesse, H. Nussbaumer, and J. P. Swings, Mon. Not. R. Astron. Soc. 151, 239 (1971).

20. M. W. Smith and W. L. Wiese, J. Phys. Chem. Ref. Data 2, 85 (1973).

21. Z. B. Rudzikas and A. P. Yutsis, Litov. Fiz. Sb. 9, 433 (1969).

22. I. S. Kychkin, Z. B. Rudzikas, and Ya. I. Vizbaraite, Litov. Fiz. Sb. 10, 47 (1970).

23. See general introduction of Ref. 20.

EXCITATION AND IONIZATION OF MODERATELY HEAVY IONS

R. H. Garstang

Joint Institute for Laboratory Astrophysics, University of
Colorado and National Bureau of Standards, and Division of Physics
and Astro-Geophysics, University of Colorado, Boulder, CO 80309

ABSTRACT

A brief review is given of some recent work on excitation, de-excitation and ionization of moderately heavy ions (iron, zinc and gallium).

INTRODUCTION

In the preceding talk at this Workshop, W. L. Wiese has discussed the situation in regard to radiative transition probabilities of iron, cobalt and nickel ions. We shall discuss mainly collisional processes; we begin with a few remarks on forbidden line transition probabilities and cross sections.

FORBIDDEN LINES

Transition probabilities have been calculated for many of the stages of ionization of iron. Perhaps the most unfortunate aspect of the situation is the complete lack of experimental checks on the calculated transition probabilities and electron collisional cross sections of the ions of interest. In one or two cases the earlier calculations of forbidden line transition probabilities have been checked: Ekberg and Edlén[1] repeated the calculations of Garstang[2] on [Fe IV], and Nussbaumer and Swings[3] repeated the calculations of Garstang[4] on magnetic dipole transition probabilities in [Fe II]. In general, reasonable agreement was obtained, with some minor differences which may in part be due to simplifying assumptions used to make the earlier calculations tractable for hand computing. However, the methods used in these calculations are basically similar, and the later work does not constitute a check of the physical correctness of the results. Having said that we hasten to add that we have no reason to doubt the general correctness of the results. Measurements on [O I] by Corney and Williams[5] are in good agreement with theory.

The only investigations of electron collisional excitation of metastable levels in the lower iron ions are those of Garstang, Robb and Rountree[6] on [Fe III] and [Fe VI] and of Nussbaumer and Storey[7] on [Fe VI]. The calculations of Nussbaumer and Storey are to be preferred for [Fe VI] because they included certain contributions from higher partial waves which were mistakenly neglected by Garstang, Robb and Rountree. The differences which resulted were roughly a factor 2, in the sense of Nussbaumer and Storey's cross sections being larger. Apart

from this the results were generally similar. Two different approximations (close coupling, distorted wave) were used, and the similarity of the results is encouraging. Once again we note that, as
in the case of the transition probabilities mentioned above, the
similarity does not provide a complete check of the physical correctness of the results. No other checks are available for these
complex ions. The criticism of the results of Garstang, Robb and
Rountree for [Fe VI] does not extend to their results for [Fe III],
for which case they did include the appropriate higher partial
waves. Many uncertainties remain in these complex calculations. We
draw attention to recent calculations on other ions, for which resonances have been shown to be important. A good example is the work
of Jackson[8] on N III, and a number of other cases have been studied.

One problem which was encountered in the work on [Fe VI] and
[Fe III] was a general lack of sensitivity of the ratios of the
emissivities of pairs of lines. In some applications the absolute
emissivity of one line is useful, but for some rougher types of
work it is useful to correlate the ratios of the intensities of
pairs of lines with the electron temperature and electron density
in the emitting region. For the stronger lines in [Fe III] and
[Fe VI] such ratios are frequently rather insensitive to the electron density and temperature; ratios which are sensitive tend to
involve at least one relatively weak transition which is correspondingly hard to observe, if observable at all, in astronomical
objects.

Perhaps the most important calculation which has still to be
undertaken in this type of work is the calculation of electron
collision excitation cross sections for the metastable levels of
Fe II. Until this becomes possible we cannot say whether there are
sensitive line pair ratios in [Fe II] which would be useful for the
diagnostics of low density objects, and we cannot produce reliable
models of such objects to predict the [Fe II] line intensities.

One possibility which has only recently been considered is the
two-photon de-excitation of metastable energy levels. Garstang[9]
pointed out that in [O I] the ^1S-^1D transition can arise from a
two-photon transition as well as from electric quadrupole radiation. The former yields a continuum, and so would be missed when
making line intensity observations. Thus line intensity observations would not give a true total radiative de-excitation rate.
Garstang's investigation showed that such two-photon processes are
indeed negligible compared with electric quadrupole radiation in
[O I] and in isoelectronic ions. Garstang[10] has recently investigated the probability of the two-photon de-excitation 4s-3d in the
Fe VIII ion, and it is also negligible in comparison with electric
quadrupole radiation. On the basis of this rather limited data it
would seem that two-photon de-excitation of metastable levels is
not likely to be important except for the hydrogen and helium isoelectronic sequences.

PERMITTED LINES

Little information is available on the cross sections for
electron collisional excitation and ionization of moderately heavy
ions, and studies have yet to be made for many neutral atoms. Only
a few groups in the world are doing experimental work, and much
effort is required to do a single set of cross-section measurements. For the excitation of ions cross-sections measurements have
been performed for only about a dozen elements in the Periodic
Table, and all but two of these measurements referred to singly-
ionized atoms. Rate coefficients have been measured for about
twenty ions, most of which were multiply-ionized. Ionization cross
sections are in a similar situation except that there exist a
number of additional relative measurements for ions of elements
such as barium and mercury. Theoretical calculations have been
performed for some cases. Attempts have been made to provide
approximate predictions of ionization cross sections (Lotz[11,12]),
and an assessment of the reliability of such predictions would be
very useful.

Work has been in progress at JILA on a number of cross-section
measurements. Space will permit only a brief description of this
work, which has been performed by Dr. G. H. Dunn and colleagues.
The method of crossed beams has been used. A beam of ions from an
ion source passes through the apparatus in the x direction and is
impacted by a beam of electrons traveling in the y direction. An
ion which has been excited by electron collision subsequently decays to its ground state again by the emission of a photon, and the
photons which are emitted in this way are recorded by a detector on
the z axis. The ion beam is also collected and may be charge analyzed to determine ionization cross sections.

An important measurement on the excitation of the 2p ^{3}P level
of Li$^+$ was completed by Rogers, Olsen and Dunn.[13] This measurement
has the distinction of being the first experimental measurement for
a positive ion of an electron collisional excitation involving a
change of spin. The dominant process is collisional excitation
from 1s^2 ^{1}S to 2p ^{3}P followed by radiation from 2p ^{3}P to 2s ^{3}S.
Interpretation of the observations is rendered more complex because
substantial cascading occurs from 3d ^{3}D and 3s ^{3}S to 2p ^{3}P. The
cross section at threshold is about 2×10^{-18} cm^2. The maximum
cross section ocurs at an electron energy of about 70 eV, and is
also about 2×10^{-18} cm^2. The observations are consistent with the
theoretically predicted finite cross section at threshold, i.e.,
with an infinitely sharp onset at threshold. The resolution obtained, which is mainly determined by the spread of energies in the
electron beam, is about 1.0 eV. There is evidence of structure in
the near-threshold cross section which may be ascribed to resonances
with doubly-excited states of neutral lithium. Above an electron
energy of 86 eV the cross section is in good agreement with the
theoretically predicted E^{-3} law. Comparisons with theoretical

calculations by several authors indicated agreement to about a factor 2. The work involved in obtaining measurements of the small cross section at higher energies (over 100 eV) is considerable, one run often amounting to 80 hours of recording.

Two other positive ions for which excitation cross sections have been measured are C^{+3} and N^{+4}. Taylor, Gregory, Dunn, Phaneuf and Crandall[14] have measured the 2s → 2p excitation in C^{+3}. Their results agreed very closely with theoretical cross sections obtained by the close coupling method. Calculations using an effective Gaunt factor, which is often satisfactory for singly-charged ions, gave a cross section a factor 3 smaller than the experimental values. Similar measurements have been made on N^{+4} by Gregory, Dunn, Phaneuf and Crandall[15] and excellent agreement was obtained between experiment and theory. Electron-impact ionization cross sections for C^{+3} and N^{+4} were obtained by Crandall, Phaneuf and Taylor[16] for electron energies up to 500 eV. Comparisons with various theoretical calculations range from good agreement to discrepancies of a factor of 2. At the highest energies (500 eV) measured the cross sections have not yet begun to decrease as predicted by all the available theories. This may perhaps be explained by a process of inner-shell excitation followed by autoionization. Rate coefficient calculations were also performed and the results compared with measurements in plasmas. The agreement was satisfactory for C^{+3}; discrepancies of a factor of 2 exist in N^{+4}.

Finally, mention must be made of recent work at JILA by Rogers[17] on excitation and ionization of Zn^+ and Ga^+. The resonance line excitation cross sections show the expected behavior of finiteness at threshold. The Zn^+ resonance line cross-section data show evidence of structure at about 12 eV, probably due to the onset of excitation of the 4d ^{2}D term. Structure is also indicated at several higher energies. Agreement with five-state close coupling calculations is good. Effective-Gaunt-factor calculations underestimate the cross sections at energies above the near-threshold region. In Ga^+ the effective-Gaunt-factor method overestimates the threshold cross section by a factor of 2. Comparison of the ionization cross sections with predictions by the method of Lotz[12] shows that the latter predictions are too large by about a factor 2.

It is clear that much work remains to be done by any available theoretical or experimental method before we can claim that we have a reliable knowledge of the excitation and ionization cross sections for positive ions of the heavier elements. The experiments we have just described do provide benchmark measurements against which future theories may be tested. For applications in astrophysics reliance will have to be placed on the results of theoretical calculations.

ACKNOWLEDGMENT

I am much indebted to Dr. Wade T. Rogers for communicating his results to me in advance of publication, and to him and Dr. G. H. Dunn for permission to use them in my presentation at the Workshop and for permission to mention them in this written version of my talk. My own work on Fe VIII was supported in part by National Science Foundation Grant PHY79-04928.

REFERENCES

1. J. O. Ekberg and B. Edlén, Phys. Scr. 18, 107 (1978).
2. R. H. Garstang, Mon. Not. Roy. astron. Soc. 118, 572 (1958).
3. H. Nussbaumer and J.-P. Swings, Astron. Astrophys. 7, 455 (1970).
4. R. H. Garstang, Mon. Not. Roy. astron. Soc. 124, 321 (1962).
5. A. Corney and O. M. Williams, J. Phys. B 5, 686 (1972).
6. R. H. Garstang, W. D. Robb and S. P. Rountree, Astrophys. J. 222, 384 (1978).
7. H. Nussbaumer and P. J. Storey, Astron. Astrophys. 70, 37 (1978).
8. A.R.G. Jackson, Mon. Not. Roy. astron. Soc. 165, 53 (1973).
9. R. H. Garstang, Optica Pura y Aplicada 10, 151 (1977).
10. R. H. Garstang, Bull. Am. Astron. Soc. 11, 641 (1979).
11. W. Lotz, Z. Phys. 216, 241 (1968).
12. W. Lotz, Z. Phys. 220, 466 (1969).
13. W. T. Rogers, J. Ø. Olsen and G. H. Dunn, Phys. Rev. A 18, 1353 (1978).
14. P. O. Taylor, D. Gregory, G. H. Dunn, R. A. Phaneuf and D. H. Crandall, Phys. Rev. Lett. 39, 1256 (1977).
15. D. Gregory, G. H. Dunn, R. A. Phaneuf and D. H. Crandall, Phys. Rev. A 20, 410 (1979).
16. D. H. Crandall, R. A. Phaneuf and P. O. Taylor, Phys. Rev. A 18, 1911 (1978).
17. W. T. Rogers, Ph.D. Thesis, University of Colorado, 1980.

THE OPACITY OF AN EXPANDING MEDIUM

Alan H. Karp
IBM Scientific Center
1530 Page Mill Road, Palo Alto, CA 94304

ABSTRACT

Shortly after the breakout of the initial shock wave generated by the collapsing core of a supernova, the expansion of the envelope becomes nearly isotropic. Hydrodynamic models of this phase can be quite simple. One complication is that standard opacity tables do not properly account for the way the radiation interacts with the expanding gas. Since a photon is continually Doppler shifted relative to the gas it is passing through, the probability that it interacts with a spectral line is greatly enhanced. It has been shown that this enhancement can be as large as an order of magnitude in some circumstances.

This paper is divided into four parts. After reviewing the expansion opacity for a homogeneous, isotropically expanding medium, we relax the assumptions. First, we consider the expansion opacity in a homogeneous, non-isotropically expanding gas. Next, we drop the assumption of homogeneity, but restrict our attention to the case of an optically thick gas. Finally, we consider the general problem in which the photons are allowed to have an arbitrarily large mean free path.

I. INTRODUCTION

The work reported here was begun to answer a specific question. When Lasher[1] computed his models of a type I supernova envelope, he assumed that the only opacity was due to electron scattering. This assumption seemed to be reasonable because at the low densities occuring in supernova envelopes the standard[2] opacity tables indicated that little else contributed. Also, the models agree with the observations up to about 30 days after the collapse of the core.

Lasher also assumed that the gas is in LTE, and this is the assumption that presents a problem. Electron scattering is coherent and, therefore, does not thermalize the radiation. We would expect that the spectrum of the supernova would be appreciably bluer than a black body at the surface temperature. The color curve indicates that this is not the case, that the radiation escaping from the envelope has indeed been thermalized. The observations[3,4] support this view. The question is, what is source of the absorption that thermalizes the radiation?

Lasher speculated that the needed absorption could be provided by spectral lines. Consider the situation

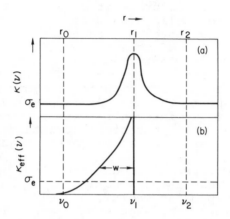

Fig. 1 - (a) A sketch of the total static opacity in the vicinity of a line. (b) A sketch showing the effective opacity due to an infinitely strong, infinitely narrow line.[5]

shown in Figure 1a. There is a photon emitted at a frequency ν_0 which, if it does not interact with a spectral line, will travel an average distance of $1/(\sigma\rho)$, a scattering mean free path, where σ is the continuous opacity and ρ is the mass density. If the gas is expanding, the photon will see the surrounding gas at a frequency ν_2 when it has traveled this distance. Now look at what happens if there is a spectral line at $\nu_0 < \nu_1 < \nu_2$. The probability that the photon will interact with the gas will be increased, and its mean free path will be correspondingly shorter. Since the opacity varies inversely with the mean free path, we can define the *expansion opacity* as

$$\kappa_{exp} = 1/\bar{x}.$$

The derivation of the expansion opacity in a homogeneous, isotropically expanding gas is reviewed in §II. The question of how non-isotropy affects the opacity is discussed in §III, and the extension to a non-homogeneous medium is given in §IV. The general problem, including that of an optically thin gas, is discussed in §V.

II. ISOTROPIC AND HOMOGENEOUS CASE

This section will follow the development of our earlier paper[5]. We wish to determine how much the opacity of a gas is increased by the effect of spectral lines if it is expanding with an inverse velocity gradient t. The quantity t is a useful measure of the velocity gradient for supernova envelopes since the gradient is given by the time since the initial collapse of the core. To put matters into perspective, our results are for $t \simeq 10$ days, which corresponds to a velocity difference of 1 km/sec over a distance of a solar radius.

Our first problem is to find the correspondence between distance traveled and Doppler shift. For isotropic expansion

$$\frac{dv}{dr} = \frac{1}{t},$$

so that

$$v_1 - v_0 = \int_{r_0}^{r_1} \frac{dr}{t} = \frac{r_1 - r_0}{t}$$

is independent of direction. Using the nonrelativistic Doppler shift equation gives

$$\frac{d\nu}{\nu} = -\frac{dv}{c} = -\frac{dr}{ct},$$

which can be integrated to give

$$r_1 - r_0 = ct \int_{\nu_0}^{\nu_1} \frac{d\nu}{\nu} = ct \ln(\nu_0/\nu_1).$$

Before calculating the expansion opacity for a realistic case, consider what happens if a photon is emitted just blueward of an infinitely narrow, infinitely strong line as illustrated in Figure 1b. Since the photon will always be absorbed by this line, its mean free path is

$$\bar{x} = r_1 - r = ct \ln(\nu/\nu_1),$$

and

$$\kappa_{eff}(\nu) = \frac{\sigma}{s \ln(\nu/\nu_1)}$$

where $s = \sigma \rho c t$ is a dimensionless number we call the *expansion parameter*. It can be shown that, for $s \gg 1$, $1/s$ is the relative Doppler shift between scatterings.

The effective width of this feature is $w = \nu/s$ which shows that the effect of the velocity gradient on the opacity will be appreciable as long as w is greater than the Doppler width of the spectral lines, i.e., as long as s is less than the speed of light divided by the thermal velocity of the ions. Since the thermal velocity is

typically 1 km/sec, this restriction is not very limiting.

If the line is not infinitely strong, there is a
finite probability that the photon will be absorbed as it
is Doppler shifted across the line. This probability is
defined in terms of the optical path length crossed by
the photon as it is Doppler shifted completely across the
line, i.e.,

$$\tau = \int_0^\infty \kappa_\nu \rho \, dr = \rho c t \int_0^\infty \frac{\kappa_\nu}{\nu} \, d\nu,$$

where

$$\kappa_\nu = \frac{\pi e^2}{mc} \frac{1}{\nu} \frac{N}{\rho} \phi_\nu = \kappa \nu \phi_\nu,$$

and the symbols have their usual meaning. We now have

$$\tau = (\sigma \rho c t) \frac{\kappa}{\sigma} \nu \int_0^\infty \frac{\phi_{\nu'}}{\nu'} \, d\nu'.$$

If we assume that the line width is small compared to the
Doppler shift experienced by the photon in traveling a
mean free path, then

$$\tau = s \frac{\kappa}{\sigma}.$$

We are now in a position to compute the effective
opacity. The mean free path is

$$\bar{\chi}(r) = \int_r^\infty (r'-r) p(r,r') \, dr',$$

where $p(r,r')$ is the probability that a photon emitted at
r is absorbed at r'. With our assumptions this integral
reduces to a sum over all lines having a frequency less
than ν. If the first line to the red of ν is very strong
and has a frequency ν', then

$$\chi(\nu) \simeq \frac{1}{\sigma \rho} \left\{ 1 - \left(\frac{\nu'}{\nu}\right)^\delta \right\}.$$

The expansion opacity can be written as

$$\kappa_{exp} = \frac{\sigma}{1-\epsilon}$$

where ϵ is called the *enhancement factor*. The results
presented below are all given in terms of this factor.

In order to demostrate the behavior of the expansion
opacity, we show in Figure 2 the ratio of the expansion
opacity to the continuous opacity for two values of the
velocity gradient and a small subset of spectral lines.
Notice that when the velocity gradient is halved, the
expansion parameter is doubled, and the individual
features get higher but narrower.

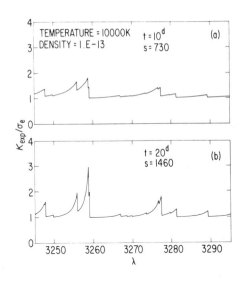

Fig. 2 - The monochromatic expansion opacity in units of the electron scattering opacity for two values of the inverse velocity gradient t. Only a small subset of all the lines was used for this example.[5]

Figure 3 shows the enhancement factor averaged over 25 Å bands for a typical temperature and density. As expected, the enhancement decreases as s gets larger. Notice that the enhancement factor is very near unity at several frequencies which implies expansion opacities orders of magnitudes larger than the scattering opacity. Another feature of interest is the dip in the enhancement near 1700 Å. If this feature is not an artifact of the line list we used[6], it might be observable. The fall-off near 500 Å is definitely due to missing lines. Since this is the soft X-ray part of the spectrum, we can make no predictions about the X-ray luminosity of supernovae until we fill in the missing lines.

This detailed spectral information is not needed for the hydrodynamic calculations, the Rosseland mean effective opacity should suffice. We can easily compute the Rosseland mean of the enhancement factor from our definition of the frequency dependent expansion opacity. Figure 4 compares the static and expansion Rosseland mean opacities for two densities and an inverse velocity gradient of 10 days. In each case the lower curve was taken from Cox and Tabor[7] while the upper curve shows the expansion opacity. Two features should be noted. The first is the large opacity enhancement at the lower density. The second is the fact that the Cox and Tabor

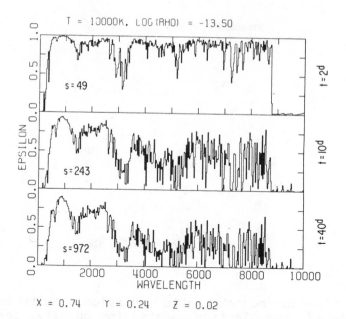

Fig. 3 - The opacity enhancement factor ε averaged over 25 Å bands for a given temperature and density and three values of the velocity gradient. (The solid portions are due to overlapping lines and have no significance.)[5]

opacities decrease with increasing densities while the expansion opacity increases. This difference may be important in determining the stability of the gas.

III. NON-ISOTROPIC FLOW

The case of isotropic flow applies to the supernova envelope only after about 5 days after the collapse of the core, and it can not be used for other objects. In this section we discuss the effect of a flow in which the velocity gradient depends on direction but in which the gas is sufficiently optically thick that neither the velocity gradient nor the physical properties of the gas change over a photon mean free path.

It can be shown that the velocity gradient in an arbitrary direction can be defined in terms of the radial velocity[8],

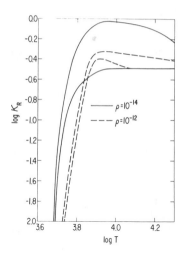

Fig. 4 - The log of the Rosseland mean opacity versus the log of the temperature for two densities. All cases were computed with and abundance of X=0.74, Y=0.28, and Z=0.02. In each case the lower curve is the static opacity[7] and the upper curve is the expansion opacity computed with an inverse velocity gradient of 10 days.[5]

$$\frac{dv}{dz} = \frac{v(r)}{r} (1+\Omega\mu^2),$$

where

$$\Omega = \frac{d(\ln v)}{d(\ln r)} - 1,$$

and μ is cosine of the angle between the radial direction and the direction of travel of the photon. Since the expansion parameter is inversely proportional to the velocity gradient,

$$\delta(\mu) = \frac{\delta}{1+\Omega\mu^2}.$$

We now have an angle dependent absorption coefficient because the integrated line strength is proportional to the expansion parameter. Since we are interested in the bulk properties of the gas, we can compute the angle averaged mean free path,

$$\bar{x} = \frac{1}{2} \int_{-1}^{1} \bar{x}(\mu) \, d\mu,$$

to be used in the calculation of the dynamics. Unfortunately, while this integral is numerically well behaved, we have been able to find an analytic solution only when $|\Omega| \ll 1$.

In order to understand the effect of the nonisotropic flow, consider the case of a photon emitted just to the blue of a strong line. As in §II we find that

$$\bar{\chi}(\mu) = \frac{1}{\sigma\rho} \left\{ 1 - \left(\frac{\nu'}{\nu}\right)^{\delta(\mu)} \right\} = \frac{1}{\sigma\rho} \{1-\exp[-\delta(\mu)(z'-z)/c t]\},$$

which reduces to

$$\bar{\chi}(\mu) \simeq \frac{z'-z}{1+\Omega\mu^z},$$

when ν' is sufficiently close to ν. The mean free path becomes very large at some angles when $\Omega < -1$ because the velocity gradient vanishes in these directions, and we neglect the continuous opacity in this approximation. Figure 5 shows the ratio of the angle averaged mean free path in terms of the isotropic mean free path as a

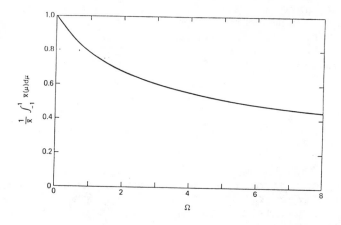

Fig. 5 - Variation of the angle averaged mean free path *versus* the anisotropy parameter Ω. The ordinate is in units of the mean free path for the isotropic case.

function of Ω in this approximation. Even for rather large anisotropies, the change is only about a factor of 2.

III. NON-HOMOGENEOUS MEDIUM

In this section we retain the assumption that the flow is isotropic and consider the effect of spatial variations. Since the envelope is assumed to be spherically symmetric, the variations are assumed to be

confined to the radial direction. Because we assume the
velocity gradient is a constant, we have the same relationship between the distance a photon travels and its
Doppler shift relative to the gas that we derived in §II.
However, as in the preceding section, the expansion
parameter becomes a function of the angle between the
direction the photon is traveling and a radius vector.
In addition, it also becomes a function of radial position due to the variation of density and electron scattering opacity.

Consider the situation shown in Figure 6. A photon
of frequency ν_0 is emitted at point r_0 into a direction

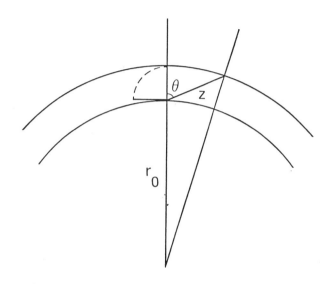

Fig. 6 - Schematic showing the geometry of a
photon moving through an inhomogeneous medium.

θ from the radial direction. Since the flow is assumed
to be isotropic, the Doppler shift experienced by the
photon depends only on the distance it travels. However,
the properties of the material it crosses depends on the
direction in which it is traveling. For example, a
photon traveling outward along a radius vector will
change its distance from the center of expansion by much
more than a photon traveling nearly perpendicular to a
radius vector. If the strength of a spectral line is a
rapidly changing function of r, these two photons will
have very different mean free paths.

134

Consider the case illustrated in Figure 7. A photon of frequency ν_0 is emitted at a position r_0 into a

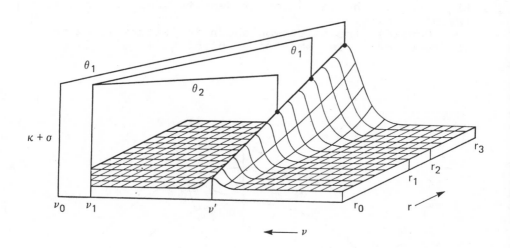

Fig. 7 - Sketch of the static opacity *versus* frequency and radial position in an inhomogeneous expanding envelope. The three straight lines represent the trajectories of photons in the frequency-radius domain.

direction θ_1. As indicated, θ_1 is small (the photon is traveling nearly radially), which means that it will travel a large radial distance before being Doppler shifted to the frequency of the line, ν'. Another photon of frequency $\nu_1 < \nu_0$ emitted at the same place and into the same direction will be Doppler shifted into the line's frequency after traveling a shorter distance. For the case shown in the figure, the line is weaker at r_2 than at r_3 so that the second photon has a longer mean free path than it would if the line were strong throughout the atmosphere. Figure 8 compares what we would see if the line were strong throughout the atmosphere with what we see for the case illustrated in Figure 7. The resulting line shape clearly depends on the variation of line strength with radial position.

Looking at Figure 8 again, we can examine the angle dependence of the expansion opacity. Consider again a photon of frequency ν_0 emitted at position r_0 but now into a direction $\theta_2 > \theta_1$. This time the photon will change its radial position only slightly while it is being

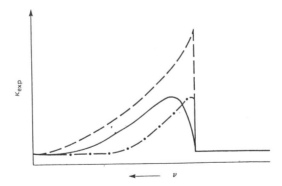

Fig. 8 - Schematic showing the effect of a radial variation of the line strength. The dashed line is for a line that is strong throughout the layer. The solid and dot-dash line are for the cases illustrated in Figure 7; the solid, for a radially moving photon; the dot-dash, for a photon moving perpendicular to the radius vector.

Doppler shifted into the frequency of the line. In this case the resulting line shape is closer to what we would expect for a homogeneous layer (Figure 8).

V. NON-ISOTROPIC AND NON-HOMOGENEOUS

It should be clear from the discussion of the previous sections that relaxing the assumptions of isotropy and homogeneity introduces serious complications. In the case of isotropic flow and a homogeneous gas, we must tabulate the expansion opacity as a function of density, temperature, and velocity gradient for each abundance. If we relax the isotropy assumption, we must also compute the expansion opacity as a function of angle. Only if the deviation from isotropy is small can the integrals over angle can be computed analytically.

The extra work that must be done if the assumption of homogeneity is relaxed is much larger. The problem is that the static opacity is a local quantity while the expansion opacity is non-local. The non-local nature means that the expansion opacity becomes model dependent. In other words, *there is no way to compute the expansion opacity in a non-homogeneous model without completely specifying the model.* Since the supernova envelope changes with time, and these changes depend on the opacity, there is no reasonable way to precompute the opacity.

This problem raises the question, What are we really after? The opacity enters into the hydrodynamic equations only through the energy and momentum equations via the flux and radiation pressure. We normally use the opacity in the diffusion approximation *because it can be computed independently of the hydrodynamics*. But we have just seen that we cannot compute the expansion opacity ahead of time if the gas is inhomogeneous. If we can't compute the opacity independently of the model, why don't we compute the flux directly?

For simplicity of discussion, we will assume plane parallel symmetry. The results are the same in spherical symmetry but the geometry is more complicated. The solution to the transfer equation is

$$I^{+}(\nu,\hbar,\mu) = \int_{\hbar}^{\infty} S(\nu',\hbar) \ \exp\{-\tau_{\nu'}(\hbar)/\mu\}\kappa(\nu')\rho \ \frac{d\hbar}{\mu},$$

where $I^{+}(\nu,\hbar,\mu)$ is the specific intensity of radiation moving outward through point $\hbar$ at an angle $\theta=\cos^{-1}\mu$ to the normal, $S(\nu,\hbar)$ is the source function, $\kappa(\nu)$ is the sum of the line and the continuous opacity, and the prime denotes that the frequency is measured in the frame of the gas. If we assume LTE, then the source function is the Planck function. A photon emitted at a frequency ν_0 will have a frequency ν' when it has traveled a continuum optical distance τ_0, where τ_0 is found by solving

$$\nu' = \nu_0 \left(1 - \mu \ \frac{\upsilon(\tau_0)}{c} \right),$$

which is accurate to order υ/c. If the line opacity differs from zero only over a very narrow range in frequency, we can write the specific intensity as[9]

$$I^{+}(\nu) = \int_{\tau}^{\tau_0} B_{\nu}e^{-\tau/\mu} \ \frac{d\hbar}{\mu} + B_{\nu}(\tau_0)e^{-\tau_0/\mu}(1-e^{-\tau/\mu})$$

$$+ \ e^{-\tau/\mu} \int_{\tau_0}^{\infty} B_{\nu}e^{-\tau/\mu} \ \frac{d\hbar}{\mu}.$$

This equation assumes a monotonic velocity law and only one line, but the generalization is obvious. A similar result can be obtained for photons moving inward.

The line shapes obtained are very similar to those computed for the isotropic, homogeneous case but the details depend on the velocity field. Figure 9 shows the specific intensity for a homogeneous model with velocity proportional to the continuum optical depth for two line strengths. Figure 10 shows the same case but with a line that varies linearly with optical depth. These figures

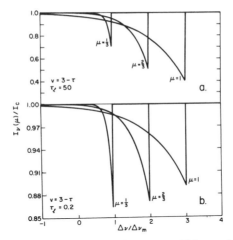

Fig. 9 - Specific intensity profiles for lines of different strengths formed in a plane parallel layer with an expanding, accelerating flow.[9]

were computed for the radiation leaving the surface. The net flux for an interior point can be obtained by subtracting the inward-going radiation from the outward-going radiation.

VI. SUMMARY AND CONCLUSIONS

It has been shown that the expansion opacity can be much larger than the conventional static opacity[5]. The problem is how best to calculate the expansion opacity if the assumptions of isotropy and homogeneity are not valid. We can look at the problem in the following way.

1. If the gas is homogeneous and expanding isotropically, compute the expansion opacity as a function of temperature, density, and chemical abundance.
2. If the gas is homogeneous and expanding nearly isotropically ($|\Omega| \ll 1$), compute the expansion opacity as above and use the analytic angle integration to correct the results.
3. If the gas is homogeneous and expanding anisotropically, compute the expansion opacity as above for a number of values of the anisotropy parameter Ω.
4. If the gas is inhomogeneous, compute the flux directly from the approximate solution to the transfer equation at every point in the model.

The important point here is that, when the gas is inhomogeneous, the expansion opacity becomes model dependent. In this case it is just as easy to compute

138

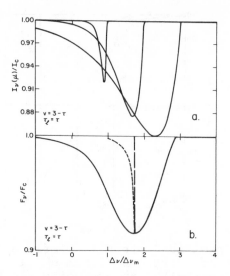

Fig. 10 - (a) Intensity profiles for the same case as Figure 9a but for a line that increases in strength with depth in the layer. (b) Flux profile for this line.[9]

the flux directly as it is to compute the expansion opacity.

I would like to thank John Greenstadt for his contributions to this paper.

REFERENCES

1. G. Lasher, *Astrophys. J.*, <u>201</u>, 194 (1975).
2. A. N. Cox and J. N. Stewart, *Astrophys. J. Suppl.*, <u>19</u>, 243 (1970).
3. R. P. Kirshner, J. B. Oke, M. V. Preston, and L. Searle, *Astrophys. J.*, <u>185</u>, 303 (1973).
4. D. Branch, in *Supernovae*, ed. D. Schramm, Reidel, Dordrecht-Holland (1977).
5. A. H. Karp, G. Lasher, K. L. Chan, and E. E. Salpeter, *Astrophys. J.*, <u>214</u>, 161, (1977).
6. R. L. Kurucz and E. Peytremann, *Smithsonian Ap. Obs. Spec. Rept.*, No. 362 (1975).
7. A. N. Cox and J. E. Tabor, *Astrophys. J. Suppl.*, <u>31</u>, 271 (1976).

DIELECTRONIC RECOMBINATION, IONIZATION EQUILIBRIUM, AND RADIATIVE EMISSION FOR ASTROPHYSICALLY ABUNDANT ELEMENTS

V. L. Jacobs and J. Davis
Plasma Radiation Group, Naval Research Laboratory
Washington, D. C. 20375

ABSTRACT

Dielectronic recombination often plays the dominant role in determining the ionization-recombination balance of multiply-charged atomic ions in low-density high-temperature plasmas. We have carried out systematic calculations of the total dielectronic re-combination rates, the corona ionization equilibrium abundances, and the radiative emission rates for all of the astrophysically important elements, including Fe and Ni. These calculations demonstrate that the inclusion of the dielectronic recombination rates produces a substantial shift in the corona equilibrium abundances of certain charge-states. This shift can have an important effect on the theo-retical prediction of the radiative energy loss rates and of the diagnostically important spectral line intensities.

INTRODUCTION

The analysis of absorption and emission lines due to atomic ions has been one of the basic methods for the determination of the physical properties of astrophysical plasmas, such as their tempera-ture and density. In addition, the energy balance in plasmas can be significantly influenced by the emission of radiation by multiply-charged ions. Usually the populations of the excited levels depart substantially from their local thermodynamic equilibrium values[1]. Consequently, the correct theoretical determination of the spectral line intensities requires a detailed treatment of many elementary collisional and radiative processes.

A substantial simplification in the theoretical determination of the spectral line intensities can be accomplished by the separa-tion of the excitation and ionization balance problems. This sepa-ration is possible only in the corona model approximation[2], which is valid for highly-charged ions in low-density plasmas. In the corona model one first determines the ionization structure assuming that electron impact ionization and autoionization following inner-shell electron impact excitation from each ground state are balanced by direct radiative and dielectronic recombination. The spectral line intensities emitted by the low-lying excited states, which are assumed to undergo spontaneous radiative decay in times that are short compared with the electron-ion collision times, are then evaluated by utilizing the corona ionization-equilibrium distri-butions of the ground states together with their electron impact excitation rates.

ISSN:0094-243X/80/630139-06$1.50 Copyright 1980 American Institute of Physics

Even though the corona ionization-recombination balance problem represents a substantial simplification of the more general collisional-radiative equilibrium equations[2], it is now recognized as a much more complex problem than was first believed. The increased complexity is the result of the discovery by Burgess[3] that the dominant recombination process for incompletely ionized ions in the important temperature region is the process of dielectronic recombination, which occurs through electron capture into doubly-excited autoionizing states. Burgess[4] was able to derive a simple formula for estimating the total dielectronic recombination rates in a low-density plasma. We have been engaged in a more detailed treatment of dielectronic recombination[5] and have discovered that the widely-used formula of Burgess can substantially overestimate the dielectronic recombination rates for certain stages of ionization because of the neglect of autoionization to excited states of the recombining ion.

Dielectronic Recombination

The dielectronic recombination of the ion $X^{+(z)}$ with residual charge Z may be described as a two-step process. First there is a radiationless capture of a plasma electron into a $n\ell$-level accompanied by the excitation $i \rightarrow j$ of the recombining ion core

$$X^{+(z)}(i) + e^- \; (\epsilon_i \ell_i) \rightarrow X^{+(z-1)} \; (j, \; n\ell). \tag{1}$$

In the corona model the initial state i is assumed to be the ground state g. Recombination is accomplished if, instead of autoionizing, the doubly-excited state j, $n\ell$ undergoes a stabilizing radiative transition to a final state k, $n\ell$ which occurs below the ionization threshold

$$X^{+(z-1)}(j, \; n\ell) \rightarrow X^{+(z-1)}(k, \; n\ell) + \hbar\,\omega. \tag{2}$$

The stabilizing transition predominantly occurs through an inner-electron transition because the radiative transition of the outer-electron is much less probable for the high n-values which usually play the dominant role.

The recombination rate per unit volume is obtained after multiplying the stabilizing radiative transition rate $A_r(j, n\ell \rightarrow k, \; n\ell)$ by the number density of ions in the doubly-excited state. This number density is determined by the balance between the electron capture rate $N_e C_e (i \rightarrow j, \; n\ell)$ and the total decay rates $A_a (j, \; n\ell)$ and $A_r(j, \; n\ell)$ due to all autoionization and spontaneous radiative processes. If it is assumed that all important singly-excited final states, k, $n\ell$ cascade freely to the ground state, the overall recombination rate is simply given by the sum of the contributions from each Rydberg series of doubly-excited levels, i.e.

$$R_d(i) = N_e \, N(i) \sum_{j,n\ell} \frac{C_e(i \to j,n\ell)}{A_a(j,n\ell) + A_r(j,n\ell)} \; A_r(j, \, n\ell \to k, \, n\ell), \tag{3}$$

where $N(i)$ is the initial ion density and N_e is the electron density.
Note that the recombination rate can be expressed in terms of a
density-independent two-body rate coefficient $\alpha_d(i)$

$$R_d(i) = N_e \, N(i) \, \alpha_d(i). \tag{4}$$

Autoionization to excited states of the recombining ion has not
been taken into account in previous calculations for the dielectronic
recombination rates. This additional process has been found to be
important[5] for the 3d, $n\ell$ doubly-excited states of boron-like
through neon-like ions of medium-Z elements, such as Fe and Ni. In
previous calculations it has been assumed that the most important
competing autoionization process is the true inverse of the radiationless electron capture process, which occurs from the 2p ground
state of the recombining ion. We have discovered that the most probable autoionization process is associated with the 3d → 3p $\Delta n = 0$
transition, in which the recombining ion is left in the 3p excited
state. Since capture from an excited state of the recombining ion
is not included in the corona model calculation, the inclusion of
this additional radiationless decay channel must result in a reduction of the recombination rate which is associated with the 3d→2p
stabilizing radiative transition. Even though dielectronic recombination is still dominant over direct radiative recombination in
corona ionization equilibrium, the neglect of autoionization to
excited states can lead to a substantial overestimation of the total
dielectronic recombination rates in the important temperature region.

The necessity of allowing for the additional autoionization
process greatly complicates the calculation of the total dielectronic recombination rates. The required autoionization rates for
each Rydberg series of doubly-excited states can be conveniently
obtained from a quantum-defect theory relationship, which expresses
the autoionization rates in terms of the threshold partial-wave
cross sections for the closely related electron impact excitation
process. In our calculations, we have employed accurate distorted
wave cross sections for electron impact excitation[6].

Corona Ionization Equilibrium

In corona ionization equilibrium the distribution of ions of a
given element, with atomic number Z, among the various charge states
is determined by the relationships

$$N_e \, N(z\text{-}1) \, S(z\text{-}1 \to z) = N_e \, N(z) \, \alpha(z \to z\text{-}1), \tag{5}$$

where z varies from 1 to Z. The total recombination rate coefficient

$\alpha(z \to z-1)$ is the sum of the direct radiative recombination rate
coefficient and the dielectronic recombination rate coefficient.
In high-temperature plasmas, such as the solar corona, dielectronic
recombination is the dominant recombination process; but direct
radiative recombination can be dominant in a recombining plasma.
$S(z-1 \to z)$ is the total electron impact ionization rate coefficient
which is the sum of the direct single-electron ionization rate
coefficient and the contribution from autoionization following
inner-shell electron excitation. Inner-shell electron excitation
and ionization can be important in an ionizing plasma.

A more general set of equations for the charge-state distributions can be written down which would take into account additional
atomic processes such as charge-exchange recombination and photoionization. In the determination of nonequilibrium ionization
structure, it may also be necessary to allow for time-variations and
transport processes.

The corona equilibrium abundances of the ions of a given element
are independent of density and are functions only of the local electron temperature. We have carried out calculations for all Fe and Ni
ions except those with 3d electrons. The ions with low ionization
potentials relative to the thermal electron energy tend to have
sharply-peaked abundance curves for which a small shift can have a
significant effect on the emission line intensity which is predicted
at a given temperature. These ions are the ones for which the
dielectronic recombination rates are typically two orders of magnitude larger than the usual direct radiative recombination rates in
the temperature region of maximum equilibrium abundance. The
temperature which is required to produce a given charge state can
be substantially underestimated by the omission of the dielectronic
recombination rates in the calculation of the ionic abundances.

Radiative Emission

The radiation emitted from a plasma as a result of electron
collisions with atomic ions may be divided into line and continuum
contributions. The line radiation may be subdivided into sponta-
neously emitted radiation from collisionally-excited states, dielectronic recombination satellite radiation which merges with the
resonance line radiation, and cascade radiation emitted following
the recombination of plasma electrons into highly-excited levels.

The rate per unit volume at which energy is radiated as a result
of the collisional excitation of the resonance lines can be expressed
in the form

$$P_Z^{(L)} = N_e N_Z \sum_{z=0}^{Z-1} \frac{N(z)}{N_Z} \sum_j \Delta E(z, g \to j) \, C(z, g \to j) \, B(z, j \to g), \quad (6)$$

where $\Delta E(z, g \to j)$ are the excitation energies, $C(z, g \to j)$ are the electron impact excitation rate coefficients, $B(z, j \to g)$ are the radiative
branching ratios, and $N(z)/N_Z$ are the corona ionization equilibrium
distributions. Reabsorption of the emitted radiation is neglected.

Departures from the corona model due to collisional deexcitation
competing with spontaneous radiative decay can be important for low-
lying excited levels. In our calculations, we have taken into
account all single-electron electric-dipole transitions for each
stage of ionization of the given element with atomic number Z.
Transitions involving changes in the principal quantum number of 0
and 1 are found to be the most important.

The power radiated in the dielectronic recombination satellites
is given by the expression which is obtained from equation (6) after
replacing the products $C(z,g \to j)$ $B(z,j \to g)$ by the dielectronic recombination rate coefficients which are associated with the important
stabilizing radiative transitions. A substantial fraction of the
dielectronic recombination satellites are unresolvable from the
resonance lines.

The radiative emission rates obtained in the corona model
approximation can be expressed in terms of density-independent two-
body rate coefficients which are functions only of the local electron temperature. The radiation which is emitted during the
stabilizing transitions in the dielectronic recombination process
can be more important than direct recombination radiation and
bremsstrahlung. However, resonance line emission is the dominant
radiative energy loss process for incompletely stripped ions in
corona equilibrium. The radiative emission rate as a function of
temperature exhibits characteristic peaks associated with the various
shells (K, L, etc.) whose heights usually increase with increasing
principal quantum number. The shift in the ionization-recombination
balance in favor of lower charge-states which is obtained from the
inclusion of dielectronic recombination tends to accentuate the
importance of the line emission as a radiative cooling mechanism.

The temperature sensitivity of the ratio of the intensities of
spectral lines arising from different stages of ionization is well-
known[1]. In the corona model approximation, this intensity ratio is
given by

$$\frac{I(z, j \to k)}{I(z', j' \to k')} = \frac{N(z)/N_Z}{N(z')N_Z} \frac{C(z, g \to j)}{C(z', g' \to j')} \frac{B(z, j \to k)}{B(z', j' \to k')} \tag{7}$$

The intensity ratio of a H-like ion resonance line to a He-like ion
resonance line increases by several orders of magnitude within a
relatively narrow temperature range. However, the usefulness of
this temperature diagnostic may be limited by the uncertainties in
the unresolvable dielectronic recombination satellite contribution
to the observed resonance line intensity.

The intensity of a resolvable dielectronic recombination
satellite relative to the intensity of the associated resonance
line provides another diagnostic of the electron temperature. The
x-ray emission spectra of high-temperature plasmas contains prominent satellites to the $1s2p\,^1P \to 1s^2\,^1S$ resonance line of He-like
ions. These satellites are the result of the transitions
$1s\,2pn\ell \to 1s^2\,n\ell$ with $n = 2$ in the corresponding Li-like ion.

In LS-coupling there are six allowed stabilizing radiative transitions, and both radiationless electron capture and inner-shell electron collisional excitation must be considered as population processes for the autoionizing levels. However, the most intense satellite line, which is produced by the $1s\ 2p^2\ ^2D \rightarrow 1s^2\ 2p\ ^2P$ transition, is formed predominantly by dielectronic recombination[7]. It follows that the relative satellite intensity in this case is independent of the ratio of the He-like and Li-like ion abundances. The accuracy of the atomic transition rates would be the only factor limiting the usefulness of this temperature diagnostic if it were not for the uncertainty in the unresolvable satellite contribution to the resonance line.

Conclusions

In corona ionization equilibrium the ionization structure is determined by assuming that electron impact ionization and autoionization following inner-shell excitation are balanced by direct radiative and dielectronic recombination. Under these conditions, the emission lines are formed primarily by electron impact excitation. In nonequilibrium plasmas, other atomic processes can play the dominant role. In the presence of intense x-ray radiation, Auger transitions following the photoionization of inner-shell electrons can substantially enhance the ionization rate. Fluorescence following inner-shell photoionization can be the dominant mechanism for the formation of x-ray emission lines. The systematic description of these additional atomic processes is the objective of our current research effort.

References

1. H. R. Griem, "Plasma Spectroscopy" (McGraw-Hill, N.Y., 1964).
2. R. W. P. McWhirter, in "Plasma Diagnostic Techniques", ed. by R. H. Huddlestone and S. L. Leonard (Academic, N.Y., 1965).
3. A. Burgess, Astrophys. J. 139, 776 (1964).
4. A. Burgess, Astrophys. J. 141, 1588 (1965).
5. V. L. Jacobs, J. Davis, P. C. Kepple, and M. Blaha, Astrophys. J. 211, 605 (1977).
6. J. Davis, P. C. Kepple, and M. Blaha, J. Quant. Spectrosc. Rad. Transf., 17, 139 (1977).
7. V. L. Jacobs and M. Blaha, Phys. Rev. A, (1980).

PHOTOIONIZATION CROSS SECTIONS CALCULATED BY MANY BODY THEORY

Hugh P. Kelly

Department of Physics, University of Virginia, Charlottesville, Va.
∘22901

ABSTRACT

The use of many body perturbation theory to calculate photoionization cross sections of atoms including electron correlations
is discussed. Results are presented for neutral argon, zinc, iron,
and chlorine. Attention is called to the effects of resonances
such as those due to 3d → 4p excitations in zinc and iron which may
affect recombination coefficients. The process of double photoionization is discussed and cross sections for neon and argon are presented.

I. INTRODUCTION

The many body perturbation theory (MBPT) of Brueckner and
Goldstone[1,2] has been very successful in accounting for the effects
of electron correlations on many atomic properties.[3] In recent
years, MBPT has been applied to the calculation of photoionization
cross sections of a number of atoms.[4] These methods have also been
used to calculate double photoionization cross sections in which
one photon is absorbed and two electrons are ejected.[5-8] In all
these calculations MBPT has been successful in accounting for electron
correlation effects including resonances. An excellent review of
different theoretical approaches to the calculation of photoionization cross sections has been given recently by Starace.[9] In Section
II we discuss the theory of MBPT as applied to photoionization cross
sections. In Section III results are presented for cross sections
of argon, zinc, iron, and chlorine. In Section IV, results are
presented for double photoionization cross sections, and Section V
contains the conclusions.

II. THEORY

A. Linked Cluster Expansion

Consider the N-electron Hamiltonian

$$H = \sum_{i=1}^{N} T_i + \sum_{i<j=1}^{N} v_{ij}, \tag{1}$$

where

$$T_i = -\nabla_i^2/2 - Z/r_i, \tag{2}$$

and $v_{ij} = e^2/r_{ij}$. Atomic units are used throughout this paper.

ISSN:0094-243X/80/630145-17$1.50

If we approximate Σv_{ij} by ΣV_i, where V_i is an arbitrary Hermitian potential, there results an approximate Hamiltonian

$$H_o = \sum_{i=1}^{N} (T_i + V_i).$$ (3)

The eigenstates of H_o satisfy

$$H_o \Phi_\alpha = E_\alpha^{(o)} \Phi_\alpha,$$ (4)

where the states Φ_α are determinants containing N single-particle solutions ϕ_n of

$$(T + V)\phi_n = \varepsilon_n \phi_n.$$ (5)

Correlation effects are included by perturbation theory with

$$H' = \sum_{i<j}^{N} v_{ij} - \sum_{i=1}^{N} V_i.$$ (6)

According to the linked cluster result of Brueckner[1] and Goldstone[2], the exact nonrelativistic solution is

$$\Psi_\alpha = \sum_{L} \left(\frac{1}{E_o - H_Q} H' \right)^n \Phi_\alpha,$$ (7)

and

$$\Delta E_\alpha = E_\alpha - E_\alpha^{(o)} = <\Phi_\alpha|H'|\Psi_\alpha>,$$ (8)

where $\sum_L$ indicates that only "linked" terms are included.

B.   Photoionization Cross Section

Consider the frequency-dependent polarizability $\alpha(\omega)$ which describes the response of the atom to an external electric field. The lowest-order contribution to $\alpha(\omega)$ from the orbital ϕ_p occupied in Ψ_α is given by

$$- \sum_{k} |<k|z|p>|^2 \left(\frac{1}{\varepsilon_p - \varepsilon_k - \omega} + \frac{1}{\varepsilon_p - \varepsilon_k + \omega} \right),$$ (9)

where the sum over k includes all excited states (i.e., those not occupied in Φ_α). If $|p>$, $|k>$ and ε_p, ε_k are exact eigenstates and eigenvalues of H, and z represents Σz_i, then Eq. (9) gives $\alpha(\omega)$ exactly.

Since $\varepsilon_p - \varepsilon_k + \omega$ may vanish, we add a small imaginary part iη and use the formula

$$\lim_{\eta \to o} (D + i\eta)^{-1} = P\ D^{-1} - i\pi\delta(D), \tag{10}$$

where P represents principal value integration. In Σ_k continuum states are included by numerical integration $(2/\pi) \int_o^\infty dk$, which assumes normalization

$$R_k \to \cos\left[kr + \delta_\ell + (q/k) \ln 2kr - \tfrac{1}{2}(\ell-1)\pi\right]/r, \tag{11}$$

as $r \to \infty$ and $V(r) \to q/r$. Bound states are included by explicit summation and use of the extrapolation

$$\left|<n|z|p>\right|^2 \to c/n^3 \tag{12}$$

for large n. Inserting Eq. (10) into Eq. (9), we find

$$\text{Im } \alpha(\omega) = (2/k)\left|<k|z|p>\right|^2, \tag{13}$$

where $k = (2\varepsilon_p + 2\omega)^{\frac{1}{2}}$.

The photoionization cross section $\sigma(\omega)$ is (in atomic units)[10]

$$\sigma(\omega) = (8\pi\ \omega/ck)\left|<k|z|p>\right|^2, \tag{14}$$

using the continuum normalization of Eq. $(11)^2$. From Eqs. (13) and (14), we obtain the usual result[11]

$$\sigma(\omega) = (4\pi\ \omega/c)\text{ Im } \alpha(\omega). \tag{15}$$

Terms in the perturbation expansion for $\alpha(\omega)$ may be represented by diagrams, and it can be shown that the many-body diagrams for Im $\alpha(\omega)$ may be factored to give the exact $<\Psi_k|\Sigma_i z_i|\Psi_\alpha>$ times its complex conjugate times corrections due to normalization diagrams.[12,13]

Diagrams contributing to $<\Psi_k|\Sigma z_i|\Psi_\alpha>$ with a $p \to k$ transition are shown in Fig. 1. The solid dot represents matrix elements of z and the cross represents interaction with $-V$. Other dashed lines represent coulomb interactions with v_{ij}. Diagrams are read from bottom to top corresponding to right to left in the mathematical expressions. Coulomb interactions below (above) the heavy dot correspond to correlations in the initial (final) state. Exchange diagrams are not shown but are understood. Diagram 1(a) represents $<k|z|p>$. Diagram 1(b) represents correlations in Ψ_k and is given by

$$\sum_{k'} <kq|v|pk'><k'|z|q>\ (\varepsilon_p - \varepsilon_{k'} + \omega)^{-1}. \tag{16}$$

148

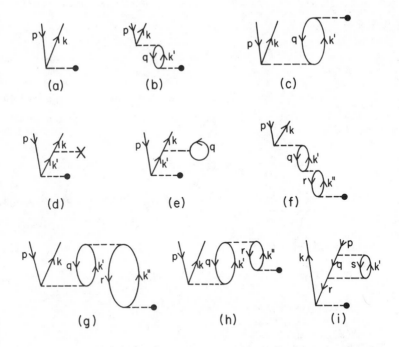

Fig. 1. Diagrams contributing to $\langle \Psi_k | \Sigma z_i | \Psi_\alpha \rangle$. The solid dot represents
a matrix element of z and the cross represents interaction with
$-V$. Other broken lines represent Coulomb interactions.
Exchange diagrams are not explicitly shown but should be
included.

Diagram 1(c), which represents ground state correlations, is given by

$$\sum_{k'} \langle q|z|k' \rangle \langle kk'|v|pq \rangle \, (\varepsilon_p + \varepsilon_q - \varepsilon_k - \varepsilon_{k'})^{-1}. \qquad (17)$$

When the Hartree-Fock potential V is used, diagrams (d) and (e) cancel.
    In calculations it is customary to evaluate both length (z) and
velocity (d/dz) forms of the matrix elements. When $|\Psi_k\rangle$ and $|\Psi_\alpha\rangle$
are exact eigenstates of H, they are related by

$$\langle \Psi_k | \Sigma z_i | \Psi_\alpha \rangle = (E_\alpha - E_f)^{-1} \langle \Psi_f | \Sigma \frac{d}{dz_i} | \Psi_\alpha \rangle. \qquad (18)$$

However, agreement of length and velocity cross sections is only a
necessary but not sufficient condition to indicate accuracy of
solutions.

C.   Choice of Potential

Although the choice of V is arbitrary, it is desirable to choose it to give Hartree-Fock orbitals occupied in Φ_α. However, this potential is not unique as can be seen by considering the potential[14-19]

$$V = R + (1-P_\alpha) \, \Omega \, (1 - P_\alpha), \tag{19}$$

where Ω is an arbitrary Hermitian operator and R is a potential which gives the N Hartree-Fock orbitals occupied in Φ_α. The operator

$$P_\alpha = \sum_{n=1}^{N} |n\rangle\langle n|, \tag{20}$$

where states $|n\rangle$ are occupied in Φ_α. When the orbital $|i\rangle$ is occupied in Φ_α, $V|i\rangle = R|i\rangle$. For an excited orbital, however, $V|i\rangle = R|i\rangle + (1 - P_\alpha) \, \Omega \, |i\rangle$.

III.   SINGLE PHOTOIONIZATION CROSS SECTIONS

A.   Argon

The ground states of atomic argon is $3s^2 3p^6 \, {}^1S$. Dipole absorption of a photon leads to many-particle states $3p^5 kd \, {}^1P$ and $3p^5 ks \, {}^1P$. If these states are represented by a linear combination of determinants, we obtain appropriate potentials for the ks and kd orbitals by setting

$$\langle 3p^5 k \, {}^1P | H' | 3p^5 k' \, {}^1P \rangle = 0, \tag{21}$$

or

$$\langle k | V | k' \rangle = \langle 3p^5 k \, {}^1P | \Sigma v_{ij} | 3p^5 k' \, {}^1P \rangle. \tag{22}$$

For ks orbitals, Eq. (19) is used to achieve orthogonality with 1s, 2s, 3s. The V of Eq. (22) results in the cancellation of Fig. 1 diagrams (d), (e), and also (b) when g belongs to the same subshell as p. Higher-order diagrams such as (f) are also cancelled by V when the hole states p, q, r, belong to the same subshell.[20,21]

The potential of Eq. (22) has a large repulsive exchange contribution. When this is not included agreement with experiment is very poor at low energies.[22]

The calculated[23] MBPT cross section for argon is shown in Fig. 2. Curves labelled HFL and HFV are the lowest order Hartree-Fock length and velocity curves with continuum orbitals calculated with V of Eq. (22). Including diagram (c) of Fig. 1 brought the length and velocity results into much closer agreement. The final curves of Fig. 2 also include the diagrams of Fig. 1 (g) and (h). The resonances are due to diagrams 1(b) and (f) with p = 3p, q = 3s, and k' = 4p, 5p, etc. Diagram 1(f) was also included with r = 3p and the imaginary part taken for the r → k" denominator. Including higher-order diagrams[23]

150

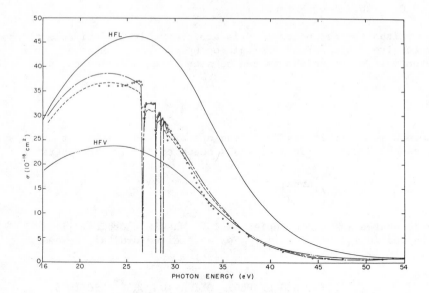

Figure 2.   Cross section for photoionization of neutral argon from
reference 23.   HFL and HFV represents Hartree-Fock
length and velocity cross sections.   Dot-dashed (dashed)
line is length (velocity) cross section including higher-
order terms.   Circles are experimental data from Madden,
Ederer, and Codling, Ref. 24.   Only the lowest 3s → np
resonances are shown.

changes 3s → np denominators from $\varepsilon_{3s}-\varepsilon_{np}+\omega$ to $\varepsilon_{3s}-\varepsilon_{np} + \Delta + i\ \Gamma_n/2$,
where Γ_n is the width of the $3s3p^6np^1P$ state and Δ is the usual
resonance shift.   Agreement with the experimental results of Madden,
Eherer, and Codling[24] is very good.   However, recent calculations[25]
using a coupled integral equations technique which sums diagrams like
1(b) and (f) to all orders gave worse agreement in the region of the
resonances than when the uncorrelated 3s → np dipole matrix elements
are used.   Including other second-order diagrams not shown in Fig. 1
tended to restore the good agreement with experiment.

   B.   Zinc

   Calculations for the $4s^2$ subshell of ZnI showed very strong
correlation effects.[26]   In Fig. 3 are shown various calculated curves
along with the experimental data of Marr and Austin.[27]   There are large
resonances in the 4s cross section due to 3d → np excitations with
n ≥ 4.   The lowest-order diagram contributing to the resonances is Fig.
1(b) with p = 4s, q = 3d, k' = np.   Higher-order diagrams may be summed
exactly geometrically to give effective widths to the resonances.

   The curves labelled NRL and NRV are length and velocity curves,
respectively, which do not include any of the resonance diagrams.   It
is striking that the Hartree-Fock length (HFL) results disagree with

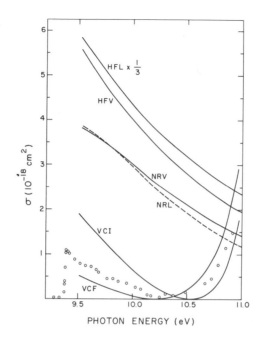

$\sigma \ (10^{-18} \ cm^2)$

PHOTON ENERGY (eV)

Figure 3. Cross section $\sigma(\omega)$ for photoionization of the 4s subshell
of ZnI near threshold. The curves HFL and HFV represent
Hartree-Fock length and velocity cross sections using frozen-
core (FC) orbitals. Curves NRL and NRV represent correlated
length and velocity cross sections omitting resonance diagrams. Curves VCF and VCI are correlated velocity cross
sections with frozen core (F) and ionic core (I) orbitals
respectively. Circles are experimental points by Marr
and Austin, Ref. 27.

experiment by almost a factor of twenty at threshold. When the
resonance diagrams were also included, velocity curves VCF and VCI
were obtained, with the corresponding length curves being quite
close. The curve labelled VCF was obtained by calculating excited
states in the field of a potential with orbitals of ZnI (i.e., frozen
core). For the curve labelled VCI, excited states were calculated in
a potential using ionic orbitals, that is, orbitals of ZnII. Therefore,
the curve VCI includes relaxation effects in an approximate fashion.

The $3d \rightarrow np$ resonance structure shows effects of spin-orbit
splitting and it is necessary to consider the interaction between
resonance states $3d^9 4s^2 np$ $^1P_1$, $^3D_1$, and $^3P_1$. The intermediate-
coupling energies and mixing coefficients for these states were calculated by the method of Wilson.[28] The calculated results are shown
in Fig. 4 and compared with experiment.[27] The absorption window
between the two large resonances is obtained only when the interaction

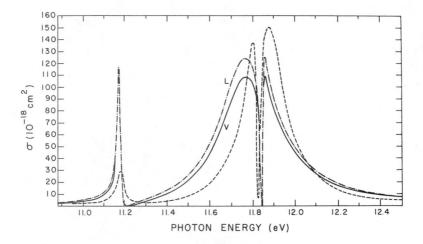

Figure 4. Photoionization cross section of ZnI 4s subshell in the $3d^94s^24p$ resonance region. Solid line (dash-dot-line) correlated velocity (length) cross section, Ref. 26. Dashed line, experimental points from Marr and Austin, Ref. 27.

between resonances is included.[26] The large 3d → np resonances are seen to lie close to threshold, and they may have a significant effect on recombination coefficients at temperatures of astrophysical interest. These types of resonances will, of course, be present for iron, cobalt and nickel which are of interest in connection with supernovae.

C. Iron

The photoionization cross section of FeI has been calculated in an approximate way by calculating orbitals to have different energies depending on m_s.[29] For example, $3d^64s^2$ is written $(3d^+)^5 (3d^-)(4s^+)(4s^-)$. Correlations among electrons are calculated by evaluating low-order diagrams. The calculations are greatly simplified over those which take proper account of LS multiplet structure. However, in a more accurate calculation, LS-coupled states should be used, and it will be interesting to carry out such calculations for iron, cobalt, nickel, and their ions. The resonances were included only by diagram 1(b) of Fig. 1 with $3d^+$ → np^+. In the results shown in Fig. 5, the resonances are actually poles since widths were not calculated, and artistic license has been used in drawing the resonance shapes.

The large resonance at 12.13 eV is due to $3d^+$ → $4p^+$ excitation and is degenerate with $4s^\pm$ → $kp^\pm$ and $3d^-$ → kf^-, kp^- excitations where k indicates a continuum state. In a proper LS picture, the $3d^+$ → $4p^+$ excitations are excited states $3d^54s^2np^5P$, ^5D, and ^5F with the core $3d^5$ coupled to give ^4P, ^4D, ^4F, or ^4G so that the $3d^+$ → $4p^+$ resonance is actually split into many smaller resonances. Consideration of

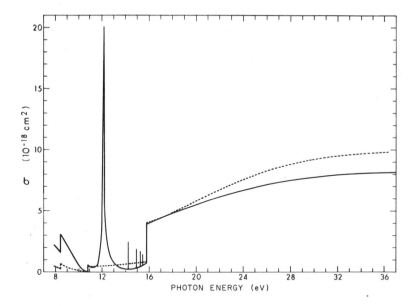

Figure 5. Total photoionization cross section of FeI calculated in
Ref. 29. Dashed line: lowest order result; solid line:
includes correlations in $4s^2$ and $3d^6$ subshells. Resonances
shown are due to $3d^+ \rightarrow np^+$ excitations as discussed in
text. Results in the vicinity of resonances have not
been accurately determined and the shape of the resonance
is very approximate.

spin-orbit splitting gives even more resonances. Indications of this
are found in a recent experimental paper by Hansen et al.[30]
    The first two thresholds shown are approximately associated
with the $3d^64s^6D$ and ^4D levels of FeII. The threshold at 10.8 eV is
associated with the $3d^54s^2$ ^6S level of FeII. The second 3d threshold
(near 15.8 eV) due to ionization of a $3d^+$ electron is associated
with the $3d^54s^2$ ^4P, ^4D, ^4F, and ^4G levels of FeII.
    The total cross section also should include contributions from
double photoionization and photoionization with excitation. Double
photoionization cross sections have so far only been calculated for a
few atoms, not including FeI. However, the cross section for photoionization with excitation to $3d^64p$ and $3d^7$ levels of FeII has been
calculated[31] and increases the single photo-cross section by
approximately 20% near threshold.

    D. Chlorine

    A number of calculations have been carried out recently on the
neutral chlorine atom as a test of theoretical methods for an open-
shell system. Random phase approximation with exchange (RPAE) calculations have been carried out by Starace and Armstrong[32] and by

Cherepkov and Chernysheva.[33] Lamoureux and Combet-Farnoux[34] have used the R-matrix approach,[35] and Brown et al. have used MBPT with LS-coupled states.[36]

Using MBPT, large contributions were found from diagrams such as Fig. 1(b) in which p and q refer to different ionic cores of ClII. For example, $p \to k$ could refer to $3p^5 \; ^2P \to 3p^4 \; ^3P \; kd^2D$ and $q \to k'$ to $3p^4 \; ^1Dk'd \; ^2D$ or $3p^4 \; ^1S \; k''d \; ^2D$. Diagonal contributions (with fixed core) are eliminated by choosing the potential V appropriately as for argon in Section IIA.

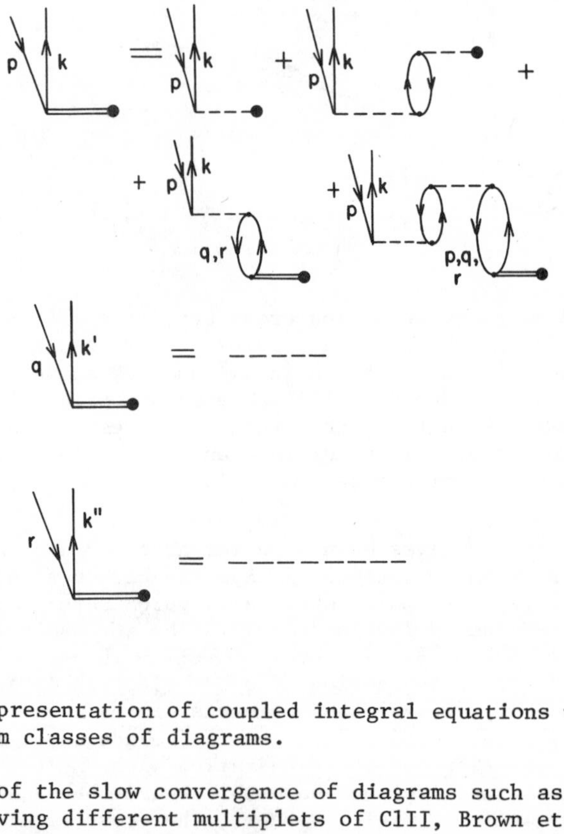

Figure 6. Representation of coupled integral equations used to sum classes of diagrams.

Because of the slow convergence of diagrams such as Fig. 1(b) and (f) involving different multiplets of ClII, Brown et al.[36] calculated these diagrams to all orders by solving the coupled integral equations indicated in Fig. 6. In these calculations the continuum was discretized at 32 points and 10 bound excited states were included. The correlated dipole matrix elements were obtained by matrix inversion. Mixing with $3s3p^6 \; ^2S$ was included in the final channel $3p^4 \; ^1D \; kd \; ^2S$.

The calculated cross sections are shown in Fig. 7. The MBPT calculation by Brown et al.[36] includes $3p \to ks$, kf and $3s \to kp$, kf cross sections. The lowest-order length result is close to that of Starace and Armstrong,[32] being 12% lower at the ^1S edge and becoming closer at higher energies. Close-coupling results by Conneely[37]

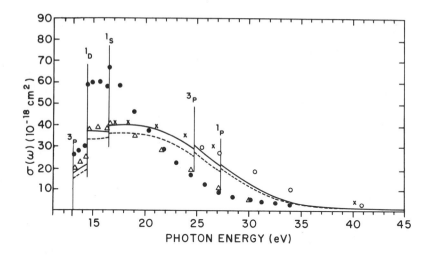

Figure 7. Calculated photoionization cross sections for ClI.
Solid (dashed) curve represents MBPT dipole length
(velocity) calculations, Ref. 36. Solid circles, RPAE
calculation by Starace and Armstrong, Ref. 32. Triangles,
RPAE calculation by Cherepkov and Chernysheva, Ref. 33.
Open circles, R-Matrix calculation by Lamoureux and
Combet-Farnoux (length), Ref. 34. Crosses, length
calculation by Conneely, Ref. 37.

(length) are also shown. The lowering of the length curve in the
final results is due both to final state mixing among ionic core
states and to initial state correlations. The MBPT and R-matrix
calculations included effects of resonances which occur before the
$3p^4$ ^1D, $3p^4$ ^1S, $3s3p^5$ ^3P, and $3s3p^5$ ^1P edges, but these are not
shown in Fig. 7.

The $3s \rightarrow kp$ cross section calculated by MBPT[36] is shown in Fig. 8.
Although the lowest order length and velocity cross sections are in
good agreement, including correlations with the $3p^5$ subshell causes a
qualitative change in the 3s cross section just as for argon.[38-40]

Relaxation effects were investigated[36] by calculating continuum
orbitals in the field of the relaxed ClIII ion. The resulting 3p cross
section is reduced by approximately 8% near the ^1S threshold and
becomes larger than the unrelaxed cross section beyond 26 eV. The
only experimental data for ClI is that of Kimura et al.[41] who made
a photoelectron measurement at 21.2 eV and determined the relative
probabilities that the resulting ClII ion was in the $3p^4$ ^3P, ^1D,
or ^1S states. These ratios were obtained in the theoretical calculations already discussed, but the best agreement with experiment was
obtained by Berkowitz and Goodman[42] who calculated geometric ratios
using squares of dipole angular factors. Of the other calculations,
the relaxed, correlated velocity results by Brown et al.[36] are in
best agreement with experiment.

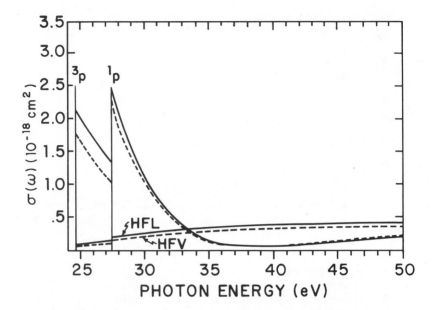

Figure 8. Total 3s subshell photoionization cross section calculated by Brown et al., Ref. 36. Upper curves are length (solid) and velocity (dashed) including correlations. Lower curves are Hartree Fock length (HFL) and velocity (HFV).

IV. DOUBLE PHOTOIONIZATION

The process of double photoionization is of considerable interest since it is due completely to electron correlations in the initial or final state. At high energies, of course, an inner shell electron may be ejected in single photoionization, and double ionization then occurs through the Auger process. There has been much experimental work on the double photoionization cross sections for rare gases[43-48], and it has been found that double photoionization contributes a significant fraction of the total cross section when energetically allowed. There have been very few calculations of this process so far: He by Byron and Joachain[49] and by Brown[50]; Ne by Chang et al.[51] and by Chang and Poe[5]; Be by Winkler[8]; and C, Ne, and A by Carter and Kelly.[6,7] All calculations but those on He used MBPT. The calculations by Carter and Kelly[6,7] used LS-coupled states which allowed a determination of LS core-structure effects.

The dipole length many-particle matrix elements may be written

$$Z(pq \rightarrow k'k) = \langle \Psi_f | \Sigma z_i | \Psi_0 \rangle, \tag{22}$$

where $|\Psi_f\rangle$ is a many-particle state with orbitals pq excited to k'k. Diagrams for $Z(pq \rightarrow k'k)$ are given in Fig. 9 with (a)-(d) being the lowest-order diagrams and (c)-(q) higher-order diagrams corresponding

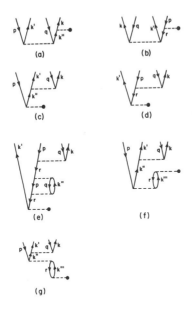

Figure 9. Diagrams contributing to the double photoionization matrix
element Z (pq → k'k). Full circles represent matrix
elements of Z. Broken lines represent Coulomb interactions.

to effects important in single photoionization calculations. Exchange
diagrams should also be included. The double photoionization cross
section is[7]

$$\sigma^{++}(\omega) = 16 \frac{\omega}{c} \int_0^{k_{max}} dk \left| \frac{Z(pq \to k'k)}{k'} \right|^2, \tag{23}$$

where continuum orbitals are normalized according to Eq. (11),

$$k' = \left[2(\epsilon_p + \epsilon_q - \frac{k^2}{2} + \omega) \right]^{\frac{1}{2}}, \tag{24}$$

and

$$k_{max} = \left[2(\epsilon_p + \epsilon_q + \omega) \right]^{\frac{1}{2}}. \tag{25}$$

In their neon and argon calculations, Carter and Kelly[7] calculated
cross sections in which an $(np)^2$ pair is ejected leaving the ion in
the np^4 ^3P, ^1D, or ^1S levels and also cross sections for nsnp pair
ejection leaving the ion $nsnp^5$ ^3P or ^1P. The angular momentum of the
outgoing k'k pair couples with the ion to give a ^1P state, and so there
is an infinite number of types of outgoing pairs. Partial cross sections calculated for argon are shown in Table I along with their
relative contributions to σ^{++}. It is surprising that 3s3p transitions

TABLE I. Argon dipole transitions.

State [a]		Contribution [b] (%)		
		ω (eV) = 89.8	100.7	239.5
$3s^2 3p^6\ {}^1S \rightarrow 3s^2 3p^4({}^3P, {}^1D, {}^1S)$	$k\,skp\,(L')^1P$	6.7	5.6	2.4
	$kpkd$	64.1	59.8	46.1
	$k\,skf$	1.8	2.4	1.7
	$kdkf$	6.7	9.4	24.2
$3s^2 3p^6\ {}^1S \rightarrow 3s3p^5({}^3P, {}^1P)$	$k'pkp\,(L')^1P$	9.8	9.7	7.9
	$k'dkd$	1.4	1.7	7.2
	$k\,skd$	9.5	11.4	10.5

[a] Notation (L') indicates that all allowed terms of the $k'k$ electron pair are to be included which are consistent with the intermediate- and final-state coupling.
[b] Geometric mean of length and velocity curves

account for more than 20% of σ^{++}. The relative contributions for neon are similar to those for argon. The calculated results for argon are compared with experiment in Fig. 10 and are surprisingly good considering that it is a low-order calculation. The results for neon[7] are also in reasonable agreement with experiment, although the discrepancy near threshold is more pronounced. This has been interpreted as due to use of a V^{N-1} potential which is asymptotically-r^{-1} rather than a V^{N-2} potential.

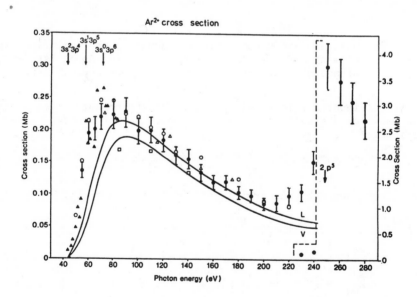

Figure 10. Double photoionization cross section for neutral argon. Solid curves are dipole length (L) and velocity (V), Carter and Kelly, Ref. 7. Experimental points: o-Holland et al., Ref. 48. Δ-Schmidt et al., Ref. 43. ⌐-Carlson, Ref. 47. ■-Lightner et al., Ref. 46. Δ-Samson and Haddad, Ref. 45. O-Wight and Van der Wief, Ref. 44.

V. CONCLUSIONS

In this paper the application of many body perturbation theory to the calculation of atomic photoionization cross sections has been discussed. Several cases have been studied which exhibit features and difficulties which should be present in calculations of photoionization cross sections and associated recombination coefficients of iron, cobalt, nickel, and their ions. The very large correlation effects on the $4s^2$ subshell cross section of Zn indicates that a similar situation should occur for the transition metal elements although these effects may be somewhat reduced due to presence of a smaller number of 3d electrons. The very important resonance effects due to 3d → 4p transitions should be included. This resonance could affect recombination coefficients and this will be studied in future work. A previous calculation for Fe was examined which shows important effects due to correlations and resonances. However, this calculation should be redone using LS-coupled states and accounting for the shapes of resonances. In the Fe calculations, no account was taken of double electron resonances due to $4s^2$ excitation, and this also should be investigated. The calculations for ClI indicate that in some cases it may be necessary to calculate to high order. This is expected, for example, for the $3d^N$ subshells of Fe, Co, Ni. However, one can calculate many terms to infinite order using the coupled integral equations approach applied to ClI. It was also seen that in accurate calculations effects of double photoionization and photoionization with excitation cannot be neglected.

I am grateful to the NSF which has supported this work.

REFERENCES

1. K. A. Brueckner, Phys. Rev. 97, 1353 (1955); The Many-Body Problem (John Wiley, N.Y., 1959).
2. J. Goldstone, Proc. Roy. Soc. (London) A239, 267 (1957).
3. H. P. Kelly, Atomic Physics 2 (ed. P. G. H. Sandars, Plenum, London, 1971), p. 227.
4. H. P. Kelly, Computer Phys. Commun. 17, 99 (1979).
5. T. N. Chang and R. T. Poe, Phys. Rev. A12, 1432 (1975).
6. S. L. Carter and H. P. Kelly, J. Phys. B9, 1887 (1976).
7. S. L. Carter and H. P. Kelly, Phys. Rev. A16, 1525 (1977).
8. P. Winkler, J. Phys. B10, L693 (1977).
9. A. F. Starace, Handbuch der Physik 31 (to be published).
10. H. A. Bethe and E. E. Salpeter, Quantum Mechanics of One and Two-Electron Atoms (Academic, New York, 1957).
11. U. Fano and J. W. Cooper, Rev. Mod. Phys. 40, 441 (1968).
12. H. P. Kelly and A. Ron, Phys. Rev. A5, 168 (1972).
13. H. P. Kelly, Advan. Chem. Phys. 14, 129 (1969).
14. L. M. Frantz, R. L. Mills, R. G. Newton, and A. M. Sessler, Phys. Rev. Lett. 1, 340 (1958).
15. B. A. Lippman, M. H. Mittleman, and K. M. Watson, Phys. Rev. 116, 920 (1959).
16. R. T. Pu and E. S. Chang, Phys. Rev. 151, 31 (1966).
17. H. J. Silverstone and M. L. Yin, J. Chem. Phys. 49, 2026 (1968).
18. S. Huzuraga and C. Arnau, Phys. Rev. A1, 1285 (1970).
19. R. L. Chase, H. P. Kelly, and H. S. Kohler, Phys. Rev. A3, 1550 (1971).
20. M. Ya, Amus'ya, N. A. Cherepkov, and L. V. Chernysheva, Zh. Eksp. Teor. Fiz. 60, 160 (1971) [Sov. Phys. JETP 33, 90 (1971)].
21. T. Ishihara and R. T. Poe, Phys. Rev. A6, 111 (1972).
22. R. L. Simons and H. P. Kelly, unpublished.
23. H. P. Kelly and R. L. Simons, Phys. Rev. Lett. 30, 529 (1973).
24. R. P. Madden, D. L. Ederer, and K. Codling, Phys. Rev. 177, 136 (1969).
25. E. R. Brown, S. L. Carter, and H. P. Kelly, unpublished.
26. A. W. Fliflet and H. P. Kelly, Phys. Rev. A10, 508 (1974).
27. G. V. Marr and J. M. Austin, J. Phys. B2, 168 (1972).
28. M. Wilson, J. Phys. B1, 736 (1966).
29. H. P. Kelly, Phys. Rev. A5, 168 (1972).
30. J. E. Hansen, B. Ziegenbein, R. Lincke, and H. P. Kelly, J. Phys. B10, 37 (1977).
31. H. P. Kelly, Phys. Rev. A6, 1048 (1972).
32. A. F. Starace and L. Armstrong, Jr., Phys. Rev. A13, 1850 (1976).
33. N. A. Cherepkov and L. V. Chernysheva, Phys. Lett. 60A, 103 (1977).
34. M. Lamoureux and F. Combet-Farnoux, J. de Physique 40, 545 (1979).
35. P. G. Burke and W. D. Robb, Advan. in At. and Molec. Phys. 11, 144 (1975).
36. E. R. Brown, S. L. Carter, and H. P. Kelly, Phys. Rev. A (to be published).
37. M. J. Conneely, Ph.D. Thesis, London University 1969 (unpublished).
38. M. Ya Amusia, V. K. Ivanov, N. A. Cherepkov, and L. V. Chernysheva, Phys. Lett. 40A, 361 (1972).

39. P. G. Burke and K. T. Taylor, J. Phys. B8, 2620 (1975).

40. J. A. R. Samson, Handbuch der Physik V. 31 (to be published).

41. K. Kimura, T. Yamazaki, and Y. Achiba, Chem. Phys. Lett. 58, 104 (1978).

42. J. Berkowitz and G. L. Goodman, to be published.

43. V. Schmidt, N. Sandner, H. Kuntzemuller, P. Dhez, F. Wuilleumier, and E. Källne, Phys. Rev. A13, 1748 (1976).

44. G. R. Wight and M. J. Van der Wiel, J. Phys. B9, 1319 (1976).

45. J. A. R. Samson and G. N. Haddad, Phys. Rev. Lett. 33, 875 (1974).

46. G. S. Lightner, R. J. Van Brunt, and W. D. Whitehead, Phys. Rev. A4, 602 (1971).

47. T. A. Carlson, Phys. Rev. 156, 142 (1967).

48. D. M. P. Holland, K. Codling, J. B. West, and G. V. Marr, J. Phys. B12, 2465 (1979).

49. F. W. Byron, Jr., and C. J. Joachain, Phys. Rev. 164, 1 (1967).

50. R. L. Brown, Phys. Rev. A 1, 586 (1970).

51. T. N. Chang, T. Ishihara, and R. T. Poe, Phys. Rev. Lett. 27, 838 (1971).

SEMIEMPRICAL CALCULATION OF gf VALUES

Robert L. Kurucz*
Harvard-Smithsonian Center for Astrophysics, Cambridge, Ma. 02138

INTRODUCTION

Because of the large scatter in my gf values in comparison to laboratory values which was well demonstrated by Wiese in his presentation, I am going to spend most of my time explaining the reasons for the scatter so that you will not worry unduly about using my calculations. Then I will mention my spectrum synthesis programs in passing and finally I will talk about future calculations that will provide oscillator strength data for Ni, Co, and Fe in supernova spectra.

SCATTER

In 1972 Eric Peytremann and I calculated gf values for 1.7 million atomic lines for sequences up through Ni making use of all laboratory spectroscopic data available at that time. The list included both lines between observed levels and lines between levels that we had predicted. We used the data statistically to compute model atmospheres for early type stars.[1] For convenience the list was subsequently edited down to fewer than one million lines by eliminating the weakest lines and that list is available from me on magnetic tape. It has been used to compute over 1200 model atmospheres for effective temperatures ranging from 5500K to 50000K and for abundance from 10 times solar down to 1/1000 solar.[2,3] The subset of that list that has lines between observed energy levels has been published by Kurucz and Peytremann[4] and is also available on magnetic tape. These data have been used in some of the supernova calculations reported at this meeting.

I am going to go through a sample calculation for the Sc II $(3d^2 + 3d4s + 4s^2)$ to $(3d4p + 4s4p)$ transition array in a very simple fashion and point out where errors can enter as we go along. [This sample calculation is not repeated here because it requires[5] 50 pages of figures and tables and because it has been published.[5] The important points are presented below.]

There are two parts to the calculation of gf values, the transition integrals between LS states, and the eigenvectors that mix the levels as combinations of LS states. The transition integrals are computed from scaled-Thomas-Fermi-Dirac wavefunctions whose

*Over the years this research has been supported in part by the Smithsonian Research Foundation and NASA grants NGR-09-015-198, NSG-7054, and NSG-5182.

eigenvalues are the observed energies. This is accomplished by
radially expanding or contracting the central potential until the
correct energy is obtained. For levels that are strongly LS coupled
there is little mixing among the LS states so that the eigenvectors
are nearly unity. Then LS allowed transitions between such levels
are determined chiefly by the transition integral. For such cases
in the iron group, whether the lines are strong or weak, my calculations are generally accurate in that they agree with the reliable
laboratory measurements as well as the reliable laboratory measurements agree with each other. This also implies that the overall
scale of my transition arrays is correct regardless of the coupling
(because mixing only divides up the total into more parts), so that
if the line data are used statistically, as in a model atmosphere
calculation, they should give the correct summed opacity.

On the other hand, LS forbidden transitions such as $d^2\ ^3P$ to
$sp\ ^3P$ or $s^2\ ^1S$ to $sp\ ^3P$ can arise only through mixing so that the
gf values of these lines crucially depend on the calculation of the
eigenvector. The large scatter observed in my gf values arises from
errors in that calculation. The eigenvectors are found from a Hamiltonian matrix that is a Slater parameter expansion including configuration interaction that is evaluated using a program written by
Bob Cowan of Los Alamos[6]. The values of the Slater parameters are
found by least squares fits such that the eigenvalues of the matrix
approximate the observed energy levels. There are a large number
of problems that can affect this least squares fit and consequently
cause errors in the eigenvectors and then in the gf values.

1. A good laboratory analysis is required. In 1972 some ions
such as Co IV and Co V had not yet been analyzed, and in many others
only the lowest configurations were (and still are) known. There
are almost always missing levels in the configurations. There are
levels for which the energy is known but not the classification, so
they cannot be used in the least squares fit. There are typograph-
ical errors and there are observed levels that do not really exist.

2. Matching the observed levels to eigenvalues is a subjective
process that must be done by hand, especially when there is strong
mixing and when there are missing levels.

3. More Slater parameters can be required than it is possible
to determine by least squares fitting. One must resort to guesses,
extrapolations, interpolations, or purely theoretical calculations
that usually have systematic errors.

4. The Hamiltonian expansion can be incomplete and leave out
some important configuration interaction. This can be investigated
with purely theoretical calculations. Fortunately, the least squares
fit absorbs some of the error from the missing physics.

5. There is such a large volume of data that mistakes are in-
evitable.

The principle means of reducing the scatter and improving the semiempirical gf values is to improve the laboratory spectral analyses which are in rather poor shape, although much improved over 8 years ago. Half the lines in the ultraviolet spectra of the sun and stars are still not identified because of the incompleteness of the laboratory analyses. Because the lines affect the total opacity, the photoionization rates, and, of course, the interpretation of the spectrum, it is extremely important to many areas of astrophysics that these data be improved.

CURRENT AND FUTURE WORK

I have a project underway for incorporating all good laboratory gf measurements into the line list, but these are only a tiny fraction of the lines. As my goal is to do research on stars, not quantum mechanics, I am happy to use any reliable laboratory data or theoretical calculations that would improve my spectra. One of the main problems with this work is that the data are published in forms that are very difficult to use. I have had research assistants working for years to determine the actual wavelengths and energy levels to which the published, rounded numbers refer. It would be simple for researchers and NBS to publish the exact values and it would also help if NBS were to publish combined energy level, multiplet, and gf tables for each ion. All new spectral analyses are being incorporated into our energy level data.

I have developed programs for computing spectra that are used to determine solar and stellar properties and abundances through the analysis of observed spectra. For a given LTE or non-LTE model atmosphere the programs compute the emergent flux or the specific intensity at up to 20 angles. The spectrum can be broadened by macroturbulence and rotation, it can be transmitted through the earth's atmosphere, broadened by the instrumental profile, and it can finally be plotted together with the observed spectrum with each line labelled. In the opacity calculation the lines are radiatively, Stark, and van der Waals broadened, and they can have isotopic and hyperfine splitting, autoionization, partial redistribution, or be merged into a continuum. Lines of ions treated in non-LTE in the model can be treated in non-LTE, and highly ionized lines can be treated in the coronal approximation. The model atmosphere can have a depth-dependent doppler shift.

Through comparison of the calculated spectra to observed spectra I also test the atomic line data. This has shown the various shortcomings in the data I have described above, and it shows the large number of missing lines in the ultraviolet. Ultimately, I would like to feed back from these calculations and from measured gf values into the semiempirical calculations. For example, the gf value for a line that arises only from configuration interaction can provide information on the configuration interaction parameters.

We have a new computer at SAO and my programs for computing gf values are still being converted. I am also trying to get Cowan's version of Froese-Fischer's Hartree-Fock program running. I plan to recompute all the previous work using updated spectral analyses. I will start with Fe II, then do the rest of Fe, Co, and Ni. In the case of Fe II the work of Johansson[7] has doubled the number of classified levels to 576 so that there will be a huge increase in the number of lines. His analysis should be further extended with new laboratory spectra in the ultraviolet.

I expect to extend the previous calculations to higher stages of ionization and to heavier elements. The programs are also being set up to produce lifetimes for each level and damping constants for each line. As mentioned above, I would also like to iterate to improve the calculation through comparison with laboratory measurements and observations.

REFERENCES

1. R. L. Kurucz, E. Peytremann, and E. H. Avrett, Blanketed Model Atmospheres for Early-Type Stars (Smithsonian Institution Press, Washington, D. C., 1975) p. 189.
2. R. L. Kurucz, Ap. J. Supp. 40, 1 (1979).
3. R. L. Kurucz, Dudley Obs. Rep. No. 14, 363 (1979).
4. R. L. Kurucz and E. Peytremann, Smithsonian Astrophys. Obs. Special Rep. No. 362 (1975).
5. R. L. Kurucz, Smithsonian Astrophys. Obs. Special Rep. No. 351 (1973).
6. R. D. Cowan, J. O. S. A. 58, 808 (1968).
7. S. Johansson, Phys. Scripta 18, 217 (1978).

REPORTS OF WORKSHOP WORKING GROUPS

INTRODUCTION

Four workshop working group meetings were held and the reports as summarized by the Chairpersons are given below. However, the atomic physics and spectroscopic data needs are not all discussed in these workshop reports. Since the authors had an opportunity to reflect further on the problems after the workshop, some of the data needs are discussed in the tutorial paper summaries. In particular the paper by A. E. S. Green should be consulted for data needs of Charged Particle Energy Deposition.

Workshop A: Requirements for Future SN Observations: γ-Ray, X-Ray UV, Visible, IR
        E. Margaret Burbidge
        Center for Astrophysics and Space Sciences
        University of California, San Diego

Workshop B: Spectroscopic Data Needs for the First Five Spectra of Fe, Co, and Ni
        W. L. Wiese
        National Bureau of Standards
        Washington, D. C. 20234

Workshop C: Recombination Rates and Recombination Spectra
        V. L. Jacobs
        Plasma Radiation Group, Naval Research Laboratory
        Washington, D. C. 20375

Workshop D: Atomic Physics and Spectroscopic Data Needs for Improved Hydrodynamic Predictions of Composition, Temperatures, and Densities of SN Envelopes
        Roger A. Chevalier
        Department of Astronomy
        University of Virginia
        Charlottesville, Virginia 22903

WORKSHOP A: REQUIREMENTS FOR
FUTURE SN OBSERVATIONS: γ-RAY, X-RAY
UV, VISIBLE, IR

INTRODUCTION

With the help especially of Dr. Robert Kirshner, who provided
summaries of the major points that emerged during the presentation and
discussion of the scientific papers, and also with the help of the
others attending this particular workshop session, I have put together
the following outline of where we stand now with regard to the observations and their interpretation, where additional (especially coordinated) observations are needed, and where new data from atomic physics
and spectroscopy are needed for interpretation of the observations.

INTERPRETATION OF OPTICAL SPECTRA

It appears that there is now a fairly good and consistent interpretation of the late-time spectra of Type I supernovae, in terms
of blends of [FeII] and [FeIII] features. We do need cross-sections
for iron-peak elements (see workshop session by Wiese). D. Branch
discussed earlier line formation in terms of the scattering of permitted Fe lines in Type I supernovae. Type II supernovae have always
presented an easier problem for interpretation, because of the presence
of H lines with clearcut broad absorption and emission profiles of the
P Cygni type.

ENERGETICS OF TYPE I SUPERNOVAE

Radioactive decay of nuclei created by nucleosynthesis during the
final collapse stages of the pre-supernova still appears to be the
best model for the energy source. Iron-peak elements are built, mainly
Ni, which decays to Co and ultimately on a longer time scale to Fe.
The observed energy output can be provided by half a solar mass of
Ni^{56}. The energy is released as γ-rays and β^+ emission. As S. Colgate
pointed out, the transparency to the γ-rays means that the energy which
is deposited can emerge without blocking. In fact, this topic really
is the realm of high-energy astrophysics—most of the energy is in
γ-rays. The exponential decay of the SN light curve can be explained
by convolution of the radioactive release with evolution of the transparency etc. of the SN. Reference was made to the work of Colgate,
Axelrod, Meyerott, and Branch.

OBSERVATIONAL DATA REQUIRED (OPTICAL, UV, IR)

A prime requirement for Type I supernovae is a long continuous
run of spectra of good resolution, i.e., about 10 Å resolution. Obser-
vatories should have contingency plans for giving priority to such
observations when a reasonably bright SN appears. Such priority is
established at the McGraw Hill Observatory, and is considered case by

case at McDonald Observatory. What is needed is a series of observations on one SN rather than scattered observations on several. Some
300-400 days after maximum, a Type I SN will be down in brightness by
5-10 magnitudes, and then 10 Å resolution, will not be possible. But
data of (say) 80 Å resolution, or multichannel data, following an
individual SN, should be obtained. All spectroscopy should, of course,
be spectrophotometry.

In addition to routine following of suitable supernovae, there are
special observational opportunities. For example, the continuum
spectrum of a SN can be used to study the interstellar medium in other
galaxies, providing information to much greater distances than can be
done otherwise. Here one needs 1½ to 2 Å resolution, covering selected wavelength regions where the interstellar lines lie.

How can observing time be obtained on the national facilities?
Time could be applied for at Kitt Peak and Cerro Tololo, for 6-month
intervals, specifying that it is a contingency program, perhaps for
Director's discretionary time or staff time. Otherwise, visitor time
is so tightly scheduled that it would be difficult to squeeze in a
continuing series of observations on a SN. Kirshner was fortunate
recently, when there was a bright Type II SN in the Virgo Cluster,
and he requested time with the International Ultraviolet Explorer (IUE).
It happened that they did have a blank shift, and were able to fit in
the observations. We need proposals for several allocated shifts to
be used as targets arise.

For the Space Telescope (ST), an emergency mode of this sort is
needed, mainly for UV observations. A long series on an individual
SN can be obtained with the ST, because of its faint observational
limit for a stellar object, e.g., a 10-12 mag. SN could be followed
down to 27 mag. Therefore, a proposal should be prepared for ST to
study bright supernovae for a long time period. Flexibility should
be built into the operations of the ST through the ST Science Institute.

Other possibilities were discussed. For example, Pennypacker:
a proposal to NASA for detecting supernovae, as a link with the Gamma
Ray Observatory (GRO). A 36-inch class automated telescope could
carry out such a photometric search program. Collaboration between
UC Berkeley and the MIRA group would be possible. Maybe half the
time on the Lick 24-inch could be used. Colgate can help with planning
such a project, searching for bright SN. It is in fact nearby SN that
are needed for GRO, which is a new start for FY 1981. It should also
be remembered that a photographic search program could also be useful.

Other optical programs were discussed, e.g., distance estimates
from expanding photospheres. Measurements of temperature and flux
give the size, so follow the angular expansion rate. Measure the
linear velocity, and this plus the angular expansion rate can give
distance. This is a much better way than treating SN as standard
candles, which they are not! One could find them at $V = 21$, out to
a redshift $z = 0.3$. It is a difficult observational program, requiring much telescope time, but Branch has done it. It could be
done out to the Coma cluster.

Colgate pointed out that if one can catch a SN before maximum brightness, and determine all three characteristic time scales of the light curve, one can obtain the Hubble constant from constancy of the first width on the curve; the redshift gives the time dilation. The ST could be used for photometry to 27th mag.

Meyerott pointed out that as well as UV observations, IR observations are important. Radiation in the lines is blackbody; measurements will be very useful.

HIGH-ENERGY DATA REQUIRED

Measurements of continuum X-rays from SN remnants are needed. Holt has seen X-ray lines of Si and Fe with the Einstein X-ray observatory. For γ-ray observations, one needs <u>early detection</u> of supernovae. An omni-directional detector is needed. The burst duration in seconds is $1/E$ (in 10eV units), i.e., at 1 keV it is 10 millisec, and at 1 MeV it is only 10 μsec. The surface is heated by the shock wave. Some 10^{48} erg may be emitted in X-rays. The time scale for measuring the <u>prompt X-rays</u> is milliseconds to hours.

Gamma-ray lines of Ni, Co, Na, Sc, Ti can be measured; the GRO sensitivity will be 10^{-5} photons cm^{-2} sec^{-1}, as reported by J. Matteson. Measurement of Fe^{60} within 20 Mpc should be easy. There will be only 5 instruments for GRO. Most instruments have a large field of view. Rapid information on new SN will be needed; it would be useful to intensify the search for SN in a given part of the sky. Matteson gave some more information: it will be possible to image to 2°; the detector is 300 cm^2; there will be 2-5 keV resolution in the MeV range, so one will be able to detect Doppler broadening and resolve a large number of gamma-ray lines.

WORKSHOP B: SPECTROSCOPIC DATA NEEDS FOR
THE FIRST FIVE SPECTRA
OF Fe, Co, AND Ni

The working group discussed the current situation on spectral
line data and data needs with respect to supernovae spectra. The
following four spectral quantities were specifically addressed: line
wavelengths; energy levels; transition probabilities for allowed lines;
and transition probabilities for forbidden lines; they will be taken
up in this order below.

As seen from a review paper at the workshop, the data situation
is much worse for transition probabilities than for wavelengths and
energy levels. Also, the differences from spectrum to spectrum
for the elements of interest are very large. With respect to supernovae work, the principal interest centers on the first five spectra
of Fe, Co and Ni, but there is also interest in some other Fe-group
elements, like Sc and Ti. Of the three "high-priority" elements, Fe,
Co and Ni, the data for the lower Fe ions are of much better quality
and are much more plentiful than those of nickel, while the cobalt
data are in the most unsatisfactory state.

The following specific needs exist:

(a) On wavelength data, knowledge of the transitions between
higher quantum states is very poor and much further work is needed.
Transitions involving principal quantum numbers n=5 and higher are
either very incompletely known or not known at all, but are needed
for comprehensive spectrum synthesis. The line identifications of
several of the spectra, especially Co I and III and Ni I and III, are
based on rather old work done in the 1940s and 1950s. Updates and
especially extensions of the spectral identifications are needed.

(b) With respect to atomic energy levels, the data situation is
very similar: again the higher levels starting at n=5 are in many
cases not known and are needed to extend the lists of forbidden lines.

(c) Reliable transition probability data for allowed lines of
the lower Fe, Co and Ni spectra are very scarce except for Fe I, as
shown in the above-mentioned review. For the lower cobalt spectra no
reliable data are available at all, with the exception of Co I.
Therefore, many additional transition probability data of high quality
are needed. While the technology to do accurate measurements is at
hand, these experiments are difficult for stages of ionization higher
than II and require well-defined pulsed high temperature sources.
Plasma sources such as small Tokamaks and pinches, which have been
recently developed, might be very useful for the purpose of such
measurements.

(d) On transition probabilities for forbidden lines, the highest
priority needs exist for data on Co I, II, and III. At present, only
very few data are available for these spectra. Aside from this
specific task, the general need on forbidden lines is to provide much
more complete line lists and transition probability data for the
lower Fe, Co and Ni spectra than presently available.

WORKSHOP C: RECOMBINATION RATES AND
RECOMBINATION SPECTRA

For the determination of the ionization-recombination balance of atomic ions in plasmas, it is necessary to evaluate the total rate of electron-ion recombination for each stage of ionization. In general, both radiative and dielectronic recombination should be taken into account. Direct radiative recombination usually occurs with the greatest probability into the ground state, giving rise to a continuous emission spectrum with a characteristic threshold. Dielectronic recombination, on the other hand, occurs primarily into excited states of the recombining ion. The radiation emitted during the stabilizing radiative transitions merges with the resonance line radiation of the recombining ion. Additional spectral line radiation is produced when the excited final state in the stabilizing transition cascades to the ground state. In order to construct a complete model of the emission spectrum, it is necessary to evaluate both the direct radiative and dielectronic recombination rates into the ground state and many excited states in each stage of ionization.

For a Maxwellian electron energy distribution, the required recombination rates can be easily obtained from the cross sections for the inverse photoionization process. If the contribution from autoionizing resonances is accounted for, both direct radiative and dielectronic recombination will be described. In particular, the usually neglected interference between these two processes, which are fundamentally the same, can be automatically included. The $n\ell$-subshell photoionization cross sections for neutral iron have been calculated using many-body perturbation-theory techniques by Kelly and Ron[1]. This calculation is expected to be particularly accurate in the threshold region because of the inclusion of correlation corrections to the usual Hartree-Fock results. Consequently, the recombination rate which can be obtained from this calculation should be very accurate at low temperature.

The $(3d)^5 (4s)^2$ n p autoionizing resonances in Fe I are expected to make a significant contribution to the recombination rate of Fe II at much lower temperatures than would usually be anticipated. This is because of their close proximity to the first ionization threshold. It should be noted that these resonances are the result of inner-shell electron excitation and are not the doubly-excited autoionizing levels usually involved in the dielectronic recombination process. Nevertheless, recombination of Fe II through such resonances can be considered as dielectronic recombination.

REFERENCE

1.   H. P. Kelly and A. Ron, Phys. Rev. A5, 168 (1971).

WORKSHOP D: ATOMIC PHYSICS AND SPECTROSCOPIC DATA NEEDS
FOR IMPROVED HYDRODYNAMIC PREDICTIONS OF
COMPOSITION, TEMPERATURES, AND DENSITIES OF SN ENVELOPES

Current hydrodynamic calculations of supernova expansion assume
that the radiation field can be characterized by a single temperature
(the blackbody temperature corresponding to the radiation energy
density). Thus, the calculations involve a mean opacity for the expanding matter. As discussed in the meeting, it is likely that a
significant fraction of the matter in a type I supernova is composed
of Ni, Co, and Fe. Near maximum light, Co would be the dominant
element. It is necessary to have extensive atomic data on these
elements over the temperature range 5×10^3 to 10^5 °K and density
range 10^{-14} to 10^{-10} g cm^{-3}. For this purpose, it would be more useful to have a large amount of data with some inaccuracies (e.g., as
computed by R. Kurucz) than to have a small amount of highly accurate
data. In this regard, it may be possible to develop general scaling
laws for the atomic parameters. Once the atomic data is available,
it will then be necessary to calculate the expansion opacity, as
described by A. Karp. The expansion effect may be particularly important for an element like Fe, which has a large number of lines.
A further problem is that at the low densities present near maximum
light, the matter is not in LTE. A detailed model atmosphere treatment of the supernova envelope may eventually be necessary.

Hydrodynamic models provide a density and temperature structure
which can then be used for line calculations. These important calculations are now being carried out by D. Branch. One problem may be
that there is radioactive energy input outside the supernova photosphere giving rise to emission features. This possibility has not
been considered in the line profile calculations of Branch. Also,
Branch has included lines of Fe II but not of Co II. It is crucial
that the atomic data on Co II lines become available so that the lines
can be included in synthetic spectrum calculations.

It now seems likely that the energy source for type I supernovae
at very late times is energetic positrons from Co^{56} decay. The main
uncertainty in the energy deposition is not in the atomic physics but
in whether the magnetic field constrains the positrons to have a
small mean free path.

With regard to observations, it is important to confirm whether
there are intrinsic variations in the light curves of SNI, as has
been claimed by the Asiago group. Thus, accurate broad-band photometry of SNI would be very useful. Observations before maximum light
would be helpful in distinguishing between various hydrodynamic models,
indicating the need for automated supermovae searches.

AIP Conference Proceedings

		L.C. Number	ISBN
No.1	Feedback and Dynamic Control of Plasmas	70-141596	0-88318-100-2
No.2	Particles and Fields - 1971 (Rochester)	71-184662	0-88318-101-0
No.3	Thermal Expansion - 1971 (Corning)	72-76970	0-88318-102-9
No.4	Superconductivity in d-and f-Band Metals (Rochester, 1971)	74-18879	0-88318-103-7
No.5	Magnetism and Magnetic Materials - 1971 (2 parts) (Chicago)	59-2468	0-88318-104-5
No.6	Particle Physics (Irvine, 1971)	72-81239	0-88318-105-3
No.7	Exploring the History of Nuclear Physics	72-81883	0-88318-106-1
No.8	Experimental Meson Spectroscopy - 1972	72-88226	0-88318-107-X
No.9	Cyclotrons - 1972 (Vancouver)	72-92798	0-88318-108-8
No.10	Magnetism and Magnetic Materials - 1972	72-623469	0-88318-109-6
No.11	Transport Phenomena - 1973 (Brown University Conference)	73-80682	0-88318-110-X
No.12	Experiments on High Energy Particle Collisions - 1973 (Vanderbilt Conference)	73-81705	0-88318-111-8
No.13	π-π Scattering - 1973 (Tallahassee Conference)	73-81704	0-88318-112-6
No.14	Particles and Fields - 1973 (APS/DPF Berkeley)	73-91923	0-88318-113-4
No.15	High Energy Collisions - 1973 (Stony Brook)	73-92324	0-88318-114-2
No.16	Causality and Physical Theories (Wayne State University, 1973)	73-93420	0-88318-115-0
No.17	Thermal Expansion - 1973 (lake of the Ozarks)	73-94415	0-88318-116-9
No.18	Magnetism and Magnetic Materials - 1973 (2 parts) (Boston)	59-2468	0-88318-117-7
No.19	Physics and the Energy Problem - 1974 (APS Chicago)	73-94416	0-88318-118-5
No.20	Tetrahedrally Bonded Amorphous Semiconductors (Yorktown Heights, 1974)	74-80145	0-88318-119-3
No.21	Experimental Meson Spectroscopy - 1974 (Boston)	74-82628	0-88318-120-7
No.22	Neutrinos - 1974 (Philadelphia)	74-82413	0-88318-121-5
No.23	Particles and Fields - 1974 (APS/DPF Williamsburg)	74-27575	0-88318-122-3

No.	Title		
No.24	Magnetism and Magnetic Materials - 1974 (20th Annual Conference, San Francisco)	75-2647	0-88318-123-1
No.25	Efficient Use of Energy (The APS Studies on the Technical Aspects of the More Efficient Use of Energy)	75-18227	0-88318-124-X
No.26	High-Energy Physics and Nuclear Structure - 1975 (Santa Fe and Los Alamos)	75-26411	0-88318-125-8
No.27	Topics in Statistical Mechanics and Biophysics: A Memorial to Julius L. Jackson (Wayne State University, 1975)	75-36309	0-88318-126-6
No.28	Physics and Our World: A Symposium in Honor of Victor F. Weisskopf (M.I.T., 1974)	76-7207	0-88318-127-4
No.29	Magnetism and Magnetic Materials - 1975 (21st Annual Conference, Philadelphia)	76-10931	0-88318-128-2
No.30	Particle Searches and Discoveries - 1976 (Vanderbilt Conference)	76-19949	0-88318-129-0
No.31	Structure and Excitations of Amorphous Solids (Williamsburg, VA., 1976)	76-22279	0-88318-130-4
No.32	Materials Technology - 1975 (APS New York Meeting)	76-27967	0-88318-131-2
No.33	Meson-Nuclear Physics - 1976 (Carnegie-Mellon Conference)	76-26811	0-88318-132-0
No.34	Magnetism and Magnetic Materials - 1976 (Joint MMM-Intermag Conference, Pittsburgh)	76-47106	0-88318-133-9
No.35	High Energy Physics with Polarized Beams and Targets (Argonne, 1976)	76-50181	0-88318-134-7
No.36	Momentum Wave Functions - 1976 (Indiana University)	77-82145	0-88318-135-5
No.37	Weak Interaction Physics - 1977 (Indiana University)	77-83344	0-88318-136-3
No.38	Workshop on New Directions in Mossbauer Spectroscopy (Argonne, 1977)	77-90635	0-88318-137-1
No.39	Physics Careers, Employment and Education (Penn State, 1977)	77-94053	0-88318-138-X
No.40	Electrical Transport and Optical Properties of Inhomogeneous Media (Ohio State University, 1977)	78-54319	0-88318-139-8
No.41	Nucleon-Nucleon Interactions - 1977 (Vancouver)	78-54249	0-88318-140-1
No.42	Higher Energy Polarized Proton Beams (Ann Arbor, 1977)	78-55682	0-88318-141-X
No.43	Particles and Fields - 1977 (APS/DPF, Argonne)	78-55683	0-88318-142-8
No.44	Future Trends in Superconductive Electronics (Charlottesville, 1978)	77-9240	0-88318-143-6